Smart Clothing

Technology and Applications

Human Factors and Ergonomics

Series Editor

Gavriel Salvendy

Professor Emeritus
School of Industrial Engineering
Purdue University

Chair Professor & Head
Dept. of Industrial Engineering
Tsinghua Univ., P.R. China

Smart Clothing
Technology and Applications

Edited by Gilsoo Cho

CRC Press
Taylor & Francis Group
Boca Raton   London   New York

CRC Press is an imprint of the
Taylor & Francis Group, an **informa** business

CRC Press
Taylor & Francis Group
6000 Broken Sound Parkway NW, Suite 300
Boca Raton, FL 33487-2742

© 2010 by Taylor and Francis Group, LLC
CRC Press is an imprint of Taylor & Francis Group, an Informa business

No claim to original U.S. Government works

Printed in the United States of America on acid-free paper
10 9 8 7 6 5 4 3 2 1

International Standard Book Number: 978-1-4200-8852-6 (Hardback)

Library of Congress Cataloging-in-Publication Data

Smart clothing : technology and applications / editor, Gilsoo Cho.
    p. cm. -- (Human factors and ergonomics)
  Includes bibliographical references and index.
  ISBN 978-1-4200-8852-6 (hardcover : alk. paper)
  1. Wearable computers. 2. Clothing and dress--Technological innovations. 3. Smart materials. 4. Ubiquitous computing. I. Cho, Gilsoo. II. Title. III. Series.

QA76.592.S63 2010
004.16--dc22
                                    2009042284

Visit the Taylor & Francis Web site at
http://www.taylorandfrancis.com

and the CRC Press Web site at
http://www.crcpress.com

Contents

Preface

The set of chapters contained here offers a unique global view for three reasons. First, they evoke the whole design cycle of smart clothes. Second, they cover applications for both the general public and professionals. Third, they dig into human aspects as well as technological aspects.

This book begins with a review and reappraisal of smart clothing by Gilsoo Cho et al., who provide a global overview by summarizing the international state of the art, identifying challenges, and evoking potential benefits of smart clothing from technological and human perspectives. Readers can thus get up to date, visualize trends, and glimpse the future.

In Chapter 2, Joohyeon Lee et al. discuss the design of technologies for smart clothing, establishing the need for methods significantly differing from traditional ones, presenting a whole theoretical design process, and providing concrete examples. Readers can relate to real cases thanks to arguments based on MP3-player jackets, photonic clothing, and bio-monitoring clothing, systems that manufacturers already commercialize though problems are by no means all solved.

In the following chapter, Yong Gu Ji and Kwangil Lee complement the discussion on design processes with a twin discussion on standardization, thus covering a critical aspect of the production and dissemination of smart clothes worldwide. They evoke trends, methods, and strategies worldwide, and detail the cases of South Korea, which is their country as well as the world leader for the production of smart clothing. Readers should value the broad scope of the information provided as well as the separate coverage of clothing and electronics.

Chapters 4 and 5 conjointly offer a view of typical enhancing components for smart clothing. Kee Sam Jeong and Sun K. Yoo present electro-textile interfaces, sensors, and actuators, and then Moo Sung Lee et al. present optical fibers. Thanks to them, the readers should understand the difficulties in choosing materials and designs that simultaneously provide targeted functions, allow a viable and elegant integration into textile and apparel, and maintain the comfort and usability of the final smart clothing in everyday life or for specific activities. As a by-product of their writing, the authors demonstrate the importance of multidisciplinary collaborations.

Reliably and efficiently exploiting combinations of components will often require particular software and hardware architectures, which will differ greatly from those existing for standard computers and multi-function cellular phones. Accordingly, Mark T. Jones and Thomas L. Martin discuss in their chapter the properties of e-textiles and propose dedicated architectures that are fault-tolerant, power-aware, and concurrently support numerous components. Although of low importance for simple cases, these aspects appear critical for complex smart clothes, and can influence their whole design.

Focusing on potential wearers, Sébastien Duval et al. explore in Chapter 7 original foundations for a global future in which smart clothes gratify human needs and match human diversity. This unique approach is theoretical and practical, clarifying trends in ubiquitous computing, testing hypotheses based on humanistic psychology in the Occident and Orient, and arguing for usefulness from birth to old age. As a result the authors propose a vision based on five key principles. Readers may consider the remarkable importance of this initiative: both meaningful starting points and clear methods are lacking to achieve projects of significant societal value, and public support remains uncertain.

In Chapter 8, Chang Gi Cho offers a deep view of shape memory materials, which possess great potential for future applications related to comfort, health, and survival, as well as aesthetics and fun, but have so far rarely been embedded into smart clothes. Readers may greatly benefit from the coverage of core aspects of shape memory materials, of a series of materials potentially very useful to design smart clothing, and of the numerous references.

In the following chapter, Daniel Ashbrook et al. sketch methods of evaluation, completing the reflections on the development cycle of smart clothes. Armed with significant first-hand experience with wearable computers, the authors provide a unique perspective. However, due to the breadth of the scope and uniqueness of their work, they could only outline the spirit in which to carry out evaluations, describe methods, and let readers be creative according to the intended wearers and smart clothes at hand. In any case, the readers should greatly benefit from this coherent approach, complementary methods, and results based on daily life as well as laboratory experiments.

Finally, Jong-Hyeok Jeon and Gilsoo Cho face the thorniest obstacle for the viability of smart clothes: the provision of energy. As a solution, they envisage creating photovoltaic textiles, textiles that absorb solar energy to transfer it as electricity to the active components. The authors first introduce the basics of solar cells, then identify milestones for the realization of photovoltaic textiles, and finally compare methods for the production of photovoltaic yarns. Readers will note that this visionary approach requires much research and development, and that success is not guaranteed. However, this first proposal may help evaluate the feasibility of the project and clarify difficulties.

I would like to thank all authors for their willingness to accept my invitation to share their pioneering efforts in this field with the readers, and for their time to prepare book chapters with their own thoughts and knowledge. Most of the authors were in the National Research Group of Technology Developments of Smart Clothing for Everyday Life sponsored by the Ministry of Knowledge Economy, Korea. I especially thank Professors Tom Martin and Mark Jones at Virginia Tech and Thad Starner at Georgia Tech for their wonderful contribution for this book. Special thanks go to Drs. Sébastien Duval and Jong-Hyeok Jeon for their participation. I am indebted to the outstanding assistance provided by all reviewers of the manuscripts. Their careful reviews and editorial suggestions improved the scientific rigor and clarity of communication in the book's chapters. I also express my gratitude to Jin Young Choi, a researcher of the smart clothing research group, for her endless devotion.

I deeply appreciate Professor Gavriel Salvendy for allowing me to edit this book in the Human Factors Book Series. Finally, I want to express my special gratitude to CRC Press for publishing this book.

Gilsoo Cho
Yonsei University
Seoul, Korea

About the Editor

Dr. Gilsoo Cho has been a professor in the Department of Clothing and Textiles at Yonsei University, Seoul, Korea, since 1984. She earned her B.S. and M.S. in clothing and textiles at Seoul National University in 1978 and 1980, respectively, and her Ph.D. in clothing and textiles at Virginia Tech in 1984.

Professor Cho currently focuses her research on the development of smart textiles and clothing. She is one of the Korean pioneers in the field. She successfully mentored 20 masters students and 7 doctoral students on diverse aspects of textile and apparel science, and has published approximately 90 articles during the last 10 years. In addition, she led various research projects, notably a 5-year project for the "technological development of smart-wear for future daily life" funded by the Korean Ministry of Knowledge Economics until 2009. She has worked with scholars from several leading universities worldwide as well as partners from Korean industrial companies.

Professor Cho has been a member of the Human Factors and Ergonomics Society since 2005, and has served on the editorial board of *Fibers and Polymers* since 2000 and is currently serving as an associate editor of the journal. She has obtained 10 patents covering topics as diverse as switches in fabrics, simulations for fabric sounds, and photovoltaic yarns. She has appeared in *Marquis Who's Who* both in science and business since 2003. She was recognized as one of the top 100 scientists in 2005 by the International Biographical Center, and received an award from the Korean Federation of Science and Technology Societies in the same year.

More information about Dr. Cho is available online at:
http://web.yonsei.ac.kr/gscho/eng/index.htm.

List of Contributors

Daniel Ashbrook
School of Interactive Computing
Georgia Institute of Technology
Atlanta, Georgia

Chang Gi Cho
Department of Fiber and Polymer
   Engineering
Hanyang University
Seoul, Korea

Gilsoo Cho
Department of Clothing and Textiles
College of Human Ecology
Yonsei University
Seoul, Korea

Ha-Kyung Cho
Department of Clothing and Textiles
College of Human Ecology
Yonsei University
Seoul, Korea

Hyun-Seung Cho
Department of Clothing and Textiles
Research Institute of Clothing and
   Textile Sciences
College of Human Ecology
Yonsei University
Seoul, Korea

Jayoung Cho
Korea Sewing Technology Institute
Daegu, Korea

James Clawson
School of Interactive Computing
Georgia Institute of Technology
Atlanta, Georgia

Sébastien Duval
Information Systems Architecture
   Science Research Division
National Institute of Informatics
Tokyo, Japan

Hiromichi Hashizume
Information Systems Architecture
   Science Research Division
National Institute of Informatics
Tokyo, Japan

Christian Hoareau
Information Systems Architecture
   Science Research Division
National Institute of Informatics
Tokyo, Japan

Jong-Hyeok Jeon
Department of Electrical and Computer
   Engineering
Birck Nanotechnology Center
Purdue University
West Lafayette, Indiana

Kee Sam Jeong
Department of Medical Information
   Systems
Yongin Songdam College
Kyeongki, Korea

Yong Gu Ji
Department of Information and
   Industrial Engineering
Yonsei University
Seoul, Korea

Mark T. Jones
Department of Electrical and Computer
   Engineering
Virginia Tech
Blacksburg, Virginia

Min-Sun Kim
Textile Fusion Technology R&D
    Department
Fusion Technology Division
Korea Institute of Industrial Technology
Ansan, Korea

Joohyeon Lee
Department of Clothing and Textiles
College of Human Ecology
Yonsei University
Seoul, Korea

Kwangil Lee
Department of Information and
    Industrial Engineering
Yonsei University
Seoul, Korea

Moo Sung Lee
School of Applied Chemical
    Engineering
College of Engineering
Chonnam National University
Gwangju, Korea

Seungsin Lee
Department of Clothing and Textiles
College of Human Ecology
Yonsei University
Seoul, Korea

Young-Jin Lee
Department of Clothing and Textiles
Research Institute of Clothing and
    Textile Sciences
College of Human Ecology
Yonsei University
Seoul, Korea

Kent Lyons
Intel Research Santa Clara
Intel Corporation
Santa Clara, California

Thomas L. Martin
Department of Electrical and Computer
    Engineering
Virginia Tech
Blacksburg, Virginia

Eun Ju Park
School of Applied Chemical
    Engineering
College of Engineering
Chonnam National University
Gwangju, Korea

Thad Starner
School of Interactive Computing
Georgia Institute of Technology
Atlanta, Georgia

Sun K. Yoo
Department of Medical Engineering
Yonsei University
Seoul, Korea

1 Review and Reappraisal of Smart Clothing*

Gilsoo Cho, Seungsin Lee, and Jayoung Cho

CONTENTS

* Much of the material in this chapter was published in the *International Journal of Human-Computer Interaction*, 25, 6, 582-617, Taylor & Francis, 2009.

1.1 INTRODUCTION

Clothing is an environment that we need and use every day. Clothing is special because it is personal, comfortable, close to the body, and used almost anywhere at any time (Kirstein et al. 2005). People enjoy clothing, with pleasures associated with its selection and wearing.

There is a need for an "ambient intelligence" in which intelligent devices are integrated into the everyday surroundings and provide diverse services to everyone. As our lives become more complex, people want "ambient intelligence" to be personalized, embedded, unobtrusive, and usable any time and anywhere. Clothing would be an ideal place for intelligent systems because clothing could enhance "our capabilities without requiring any conscious thought or effort" (Mann 1996). Clothing can build a very intimate form between human–machine interaction.

Smart clothing is a "smart system" capable of sensing and communicating with environmental and the wearer's conditions and stimuli. Stimuli and responses can be in electrical, thermal, mechanical, chemical, magnetic, or other forms (Tao 2001).

Smart clothing differs from wearable computing in that smart clothing emphasizes the importance of clothing while it possesses sensing and communication capabilities (Barfield et al. 2001). Wearable computers use conventional technology to connect available electronics and attach them to clothing. The functional components are still bulky and rigid portable machines and remain as non-textile materials. While constant efforts have been made toward miniaturization of electronic components for wearable electronics, true "smart clothing" requires full textile materials for all components. People prefer to wear textiles since they are more flexible, comfortable, lightweight, robust, and washable (Kirstein et al. 2005). To be a comfortable part of the clothing, it is necessary to embed electronic functions in textiles so that both electronic functionality and textile characteristics are retained. Smart clothing should be easy to maintain and use, and washable like ordinary textiles. Therefore, combining wearable technology and clothing/textile science is essential to achieve smart clothing for real wearability.

Smart clothing will provide useful services in numerous fields such as healthcare and warfare, where smart clothes can be designed to perform certain functions and support specialized activities, or sports and leisure, with more emphasis on aesthetics and convenience.

Developing smart clothes requires multidisciplinary approaches involving textile, human, and information science. Although smart clothing has progressed in various fields, advances remain in individual fields; more comprehensive reviews should associate diverse perspectives.

Here, we provide an overview of discoveries and issues in smart clothing. First, we review recent developments in technologies. Then, we consider human aspects and applications of smart clothing. Based on the current status of smart clothing, we suggest the direction to develop smart clothing and future work.

1.2 SMART CLOTHING TECHNOLOGY

In smart textiles and clothing, the extent of intelligence can be divided into *passive smart*, *active smart*, and *very smart systems* (Tao 2001). *Passive smart systems* can only sense the environment; *active smart systems* can sense and react to the stimuli from the environment; and *very smart systems,* in addition, adapt their behavior to circumstances.

A smart clothing system comprises (1) interfaces, (2) communication components, (3) data management components, (4) energy management components, and (5) integrated circuits (Tao 2005a). An interface is a medium for transacting information between the wearer and devices or the environment. A communication links components of the clothing, transferring information and energy. Data management refers to memory and data processing. Energy management relates to energy supply and storage. Integrated circuits are miniature electronic circuits built on a semiconductor substrate.

1.2.1   INTERFACE TECHNOLOGIES

Input and output interfaces transfer information between the wearer and devices or the environment.

1.2.1.1   Input Interfaces

Buttons and keyboards are used as input interfaces and are relatively simple and easy to learn and implement in clothes (Tao 2005a). For complex tasks, more powerful input interfaces, such as speech recognition, are needed. Sensors can monitor the context, e.g., the wearer's physiological state or location. Much effort focuses on developing textile-based interfaces for smart clothing.

1.2.1.1.1   Textile-Based Buttons and Keyboards

Conductivity in textiles is essential to smart clothing since electrical conductivity provides pathways to carry information or energy for various functions (Lam Po Tang and Stylios 2006). Conductivity in textiles can be imparted at various textile stages. Conductive polymers, fibers, yarns, fabrics, embroidery, and finishing are all vital to construct smart clothes.

Textile-based buttons and keyboards are developed based on various mechanisms. The SOFTswitch (http://www.softswitch.co.uk) is an example of pressure-sensitive textile material. It consists of conductive fabrics with a thin layer of elasto-resistive composite, called a "quantum tunneling composite." The composite is an isolator that turns into a metal-like conductor when compressed, transforming mechanical pressure into electrical signals. This "touch-sensitive" material can serve as a switch or pressure sensor.

Sensory Fabric (Swallow and Thompson 2001) consists of two conductive fabric layers separated by a meshed non-conductive layer. When the material is pressed, the two conductive layers touch through the holes in the non-conductive mesh. This pressure-sensitive fabric can serve as a switch, soft keypad, and pressure sensor.

Another system uses a multi-layer structure to form a resistive touchpad. ElekTex is a laminate of five fabric layers, in which the outer and central layers are conductive but separated by insulating layers (http://www.eleksen.com). When touched, the layers are compressed and form an electronic circuit that generates positional values (X and Y) with a low-resolution pressure measurement (Z).

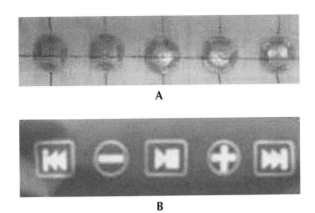

FIGURE 1.1   Switch fabric and textile-based keypad. (A) Switch fabric. (B) Textile-based keypad.

A textile-based keypad developed at the Smart Wear Research Center, Yonsei University (Figure 1.1), was fabricated using a "switch fabric." Stainless steel yarns are used as warp and filling with other types of yarn and a metal dome switch is inserted. The "switch fabric" works by contact between the conductive warp and filling yarns and the metal dome switch when compressed.

1.2.1.1.2   Textile-Based Body-Monitoring Sensors and Electrodes

Sensors measure and monitor physiological or environmental data and can act as input interfaces. Fabric-based sensors and electrodes have been developed from conductive fabrics and fiber optics.

1.2.1.1.2.1 Physiological Information   Textile sensors serve to record electrocardiograms (ECGs), respiration rates, heart rates, etc. Conventional sensors often cause problems due to their physical structure or functional requirements. For example, they may cause skin irritation either due to the adhesive or gel of conventional ECG electrodes (Catrysse et al. 2004). Textile sensors are developed to overcome these inconveniences.

Van Langenhove and Hertleer (2004) developed textile electrodes for ECG and heart rate measurements. The so-called "Textrodes" are made of stainless steel fibers and have a knitted structure, in direct contact with the skin. They were incorporated into a belt for the thorax. The textile electrodes provide accurate signals as compared with conventional electrodes, despite additional noise. This technology can help monitor patients in clinical conditions and healthcare, athletes during physical activities, professionals in extreme environmental conditions, etc.

The Smart Wear Research Center, Yonsei University, developed textile-based ECG electrodes using embroidery. Stainless steel yarns were used to embroider electrodes (Figure 1.2). The embroidered electrodes were attached to knitted shirts with spandex content from 0% to 7% to examine the effect of fabric elasticity on ECG monitoring and on wearer's comfort. The performance of embroidered electrodes will be further discussed in the Section 1.3.5.

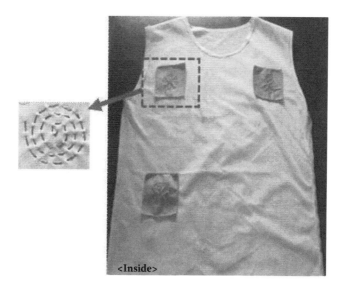

FIGURE 1.2 ECG shirt with embroidered electrodes.

In the work of Loriga et al. (2005), conductive and piezoresistive yarns were integrated in a knitted garment and used as sensors and electrodes to monitor cardiopulmonary activity. Strain fabric sensors were realized from conductive yarn as the piezoresistive domains. The fabrics exhibited piezoresistive properties in response to an external mechanical stimulus, and a voltage divider converted resistance from the fabric piezoresistive sensors into voltage. Fabric electrodes were realized with a yarn in which a stainless steel wire was coiled around a cotton-based yarn. Electrocardiogram and impedance pneumography signals were obtained from the fabric sensors and electrodes.

Catrysse et al. (2004) developed the "Respibelt," a textile sensor for measuring respiration. Made of a stainless steel yarn and knitted in a Lycra®-containing belt, it provided an adjustable stretch. The Respibelt was worn around the abdomen or thorax, and changes in circumference and length due to breathing were measured, from changes in resistance and inductance; similarly, thoracic changes in perimeter and cross-section were obtained from resistance and inductance variations.

Brady et al. (2005) integrated a foam-based pressure sensor into a garment to monitor the wearer's respiration rate. The sensor was fabricated by coating polyurethane foam with a conducting polymer, polypyrrole (PPy). The conducting polymer-coated foams were soft, compressible, and sensitive to forces from all three directions, unlike coated fabrics that work in two dimensions. The foam sensors measure chest expansion based on the compression of the foam structure between the body and the garment, whereas the conductive fabric sensors, described earlier, measure respiration rate based on the expansion and contraction of the rib cage from the stretch of the sensor.

In recent years, fiber optic technologies have attracted much attention because they offer both sensing and signal transmission. Fiber Bragg-grating (FBG) sensors are fabricated by modulating the refractive index of the core in a single-mode optic fiber to detect the wavelength-shift induced by strain or temperature change (Tian and Tao 2001; Yang, Tao, and Zhang 2001). FBG sensors contain a diffraction grid that reflects the incident light of a certain wavelength in the direction from which the light is coming. The value of this wavelength linearly relates to a possible elongation or contraction of the fiber. In this way, the Bragg sensor can function as a sensor for deformation. They have been used to monitor the structural condition of fiber-reinforced composites, concrete constructions, or other construction materials. The potential applications of FBG sensors in smart clothing include health monitoring, impact detection, shape control, and so on.

1.2.1.1.2.2 Impact/Hazard Detection   Jayaraman and coworkers (Lind et al. 1997; Park and Jayaraman 2001) used plastic optical fibers to monitor fabric damage, providing a way to detect the location of impact wounds for soldiers. In the "wearable motherboard," plastic optical fiber is integrated into the garment during the fabric production process. When a projectile, e.g., bullets or shrapnel, penetrates, broken paths in the fabrics inform about the location and degree of damage. This technology can serve for protective clothing to inform about bullet penetration, and chemical, thermal, and physical attacks. Appropriate rescue and medical treatments are then facilitated.

El-Sherif, Yuan, and MacDiarmid (2000) developed and integrated fiber optic sensors into military uniforms. The optical fibers have a chemical agent or an environmentally sensitive cladding material, which can change light propagation. The sensors can detect various battlefield hazards, such as chemicals, biological agents, and thermal variations.

1.2.1.1.2.3 Body Movement/Body Position   Farringdon et al. (1999) developed knitted stretch sensors that measure stretch from resistive changes in knitted strips. Fabric is specially knitted to combine 10-mm wide conductive threads. The wearer's movement can be monitored when the fabric is stretched and the resistance of these threads changes.

There is a growing interest in intrinsically conductive polymers (ICPs) for use in smart clothing as sensors, actuators, etc. ICPs are promising candidates for wearable systems since they possess mechanical, electrical, electronic, magnetic, and optical properties of metals. ICPs include polypyrrole (PPy) and polyaniline (PAni). Alone or mixed with conventional polymers, they can produce conductive fibers or coating materials.

De Rossi, Della Santa, and Mazzoldi (1999) reported that fabrics coated with a thin layer of conducting polymers possess strain and temperature sensing properties. Useful combinations of conducting polymers and fabrics include polypyrrole (PPy) and Lycra® because of high piezoresistive and thermoresistive coefficients with being elastic and conformable to the human body. PPy-coated Lycra® fabrics compared well with sensitive strain gauge materials and inorganic thermistors.

1.2.1.1.3 Speech Recognition (Audio Interface)

Randell and Muller (2000) developed the Shopping Jacket, in which a speech interface was integrated to assist shopping. The jacket uses location-sensing systems to alert the wearer about nearby interesting shops or to guide in a shopping mall. For interactions, a throat microphone with speech recognition software was embedded into a conventional sports blazer and provided the user input for commands and audio notes.

1.2.1.1.4 Others

Baber (2001) developed flex sensors that allow the movements of fingers to correspond to numbers, acting as input interfaces. Fitted into a glove, the sensors send digits via a microcontroller to the display when the wearer bends his or her fingers. For instance, a single bend sends 0-4, and a double bend corresponds to 5-9. The wearer can change to a control mode via a small switch mounted on the thumb. The technology offers a means of controlling devices or entering data.

1.2.1.2 Output Interfaces

An output interface is a medium by which information is presented to the wearer. Visual, auditory, and tactile interfaces are major means to transmit information from wearable systems to human.

1.2.1.2.1 Visual Interfaces

Visual displays are still the dominant output devices of conventional computing systems. To be worn on the human body as a part of smart clothing, visual displays must be compliant and conformable to the body. Flexible displays for smart clothing are under development.

Organic and polymeric light-emitting diodes (OLEDs and PLEDs) have attracted considerable attention in recent years for use in flat-panel displays where liquid crystal displays (LCDs) are the major display technology. OLEDs offer higher contrast, a higher level of brightness, a full viewing angle, and require less power than competing technologies (Tao 2005b). Flexible OLEDs, i.e., OLEDs fabricated on a flexible substrate such as plastic or metallic foil, have the advantage of being light and conformable. Conjugated polymers can provide good flexibility and mechanical properties, making them promising candidates for flexible OLEDs.

Textile-based flexible displays based on optical fibers have also been investigated. Koncar, Deflin, and Weill (2005) weaved 0.5-mm-diameter poly(methylmethacrylate) (PMMA) optical fibers with other textile yarns to construct a flexible display. The system included a small electronic device to control the light-emitting diodes (LEDs) that illuminate groups of fibers. Various weaving styles can be applied with optical fibers, including dobby and jacquard weaves.

1.2.1.2.2 Auditory Interfaces

Audio interfaces are used in portable devices and are particularly useful because they do not require one's full attention or disrupt the foreground activity.

Sawhney and Schmandt (1998) discussed techniques and issues related to the use of speech and audio in wearable interfaces. They reported designs of audio devices for wearable systems (e.g., Nomadic Radio, Radio Vest, and Soundbeam Necklace) and demonstrated a wearable audio interface utilizing various interaction techniques. The infrastructure for the concept of Nomadic Radio and design concepts for the functioning of such a system were developed.

1.2.1.2.3   Vibration (Tactile) Interface

Tactile displays are an effective tool in smart clothing because of permanent proximity to the skin. Tactile displays do not conflict with audio or visual displays and might help present information when those other displays are physically or socially inappropriate.

Tan and Pentland (1997) reported initial work on the development of a wearable tactile display. A 3×3 array of micromotors was embedded in the back of a vest, which delivered vibrational patterns to the back of the wearer. Tests with several stimulation patterns showed that the wearers could perceive directional information from the vibrational patterns. This might convey directional information for navigation guidance.

Toney et al. (2003) integrated a vibrotactile display and support electronics into a standard clothing insert, the shoulder pad, using electromagnetic motors. They proposed guidelines for design and integration of a shoulder-mounted tactile interface and showed that such a shoulder worn tactile display can make use of multiple stimulators at a low level of resolution.

Conductive polymer actuation properties have been receiving much attention in recent years. Fabrics containing conductive polymer fiber bundles that contract and relax under electrical control can be used as a tactile output interface (De Rossi, Della Santa, and Mazzoldi 1999). In the work of De Rossi, Della Santa, and Mazzoldi (1999), spun conductive polymer fibers such as polyaniline (PAni) exhibited electrochemical actuation.

1.2.1.2.4   Others

Shape memory materials (SMMs) are materials that return to a prescribed shape with the right stimuli such as heat or electrical currents. A temperature change normally stimulates SMMs and modifies their internal structure (Lam Po Tang and Stylios 2006). The two most common types of shape memory material are shape memory alloys (SMAs) and shape memory polymers (SMPs).

SMAs can exist in the form of yarns, making them comparable to textile materials. Nickel-titanium and copper-based alloys are typical examples of SMAs (Lane and Craig 2003). Winchester and Stylios (2003) spun SMA in combination with traditional fibers and created bi-component yarns, which were then developed into knitted structures. In a shirt developed by Corpo Nove (www.corponove.it), SMA was woven with traditional textile material. The shirt shortens its sleeves when the temperature increases, and creases in the fabric disappear when stimulated.

SMPs exhibit higher extensibility, superior processability, lower weight, and better hand and touch than SMAs (Lam Po Tang and Stylios 2006). SMPs include segmented polyurethane-based polymers, crosslinked poly(cyclooctene), and poly(lactic acid) and poly(vinylacetate) blends. SMPs can be extruded as fibers and used as filament yarns or spun in combination with other fibers, which can be incorporated into knitted or woven structures (Chan Vili 2007).

1.2.2 COMMUNICATION

Communication refers to information and power transfer between the components of smart clothes. We consider short-range communication as communication within a device or between two devices worn by the user; long-range communication refers to communication between two users (Tao 2005a). Diverse techniques exist for textile-based networking.

1.2.2.1 Short-Range Communication

On-body communication can be wired or wireless, including embedded wiring, infrared, and Bluetooth technology.

1.2.2.1.1 Embedded Wiring

Considerable work has been done to replace traditional wires with textile-based networks in smart clothing. Techniques include conductive fibers, yarns, fabrics, embroidery, and optical fibers.

Post and Orth (1997) built electronic circuits from various conductive textiles. Conductor lines were realized by embroidering metal fibers or weaving silk threads wrapped in thin copper foil. "Gripper snaps," a common sewing closure, connected conductive fabrics and electronics (Post et al. 2000). "Electric suspender," developed by Gorlick (1999), contains stainless steel conductors for power and data buses.

In the work of Dhawan et al. (2004), conductive threads were woven to develop woven fabric-based electrically conductive circuits. Conductive and nonconductive threads were arranged and interlaced to form woven conductive networks, and interconnects were developed at the crossover points of orthogonal conductive threads.

The Smart Wear Research Center, Yonsei University, developed textile-based transmission lines (Figure 1.3) using Teflon-coated stainless steel yarns. The conductive bands were fabricated by weaving with the stainless steel yarn as warp and polyester yarns as filling and warp. Five strands of signal transmission lines were placed at 2.54-mm intervals in the warp direction of the band, and the width of the band was set at 14 mm to ease connection with regular connectors. As illustrated in Figure 1.3, the bands were very flexible. The textile-based transmission lines network electronic components in smart clothes.

Jayaraman and coworkers (Lind et al. 1997; Park and Jayaraman 2001) developed a garment in which electrically conductive fiber and plastic optical fibers transfer information from sensors to processing units. As mentioned earlier, the garment was designed to monitor vital signs in combat personnel, detect bullet wounds, and transfer information on bullet penetration. These tasks were performed in real time by interfiber electrical connections and connectors for power and data interconnects to and from the garment.

1.2.2.1.2 Wireless Short-Range Communications

Infrared communication is common in remote controls, laptops, and digital cameras. Starner, Kirsch, and Assefa (1997) developed the Locust Swarm system, which

FIGURE 1.3   Textile-based transmission lines.

provides messaging and location information; it transferred indoor position information by infrared. Infrared requires line-of-sight but is inexpensive and suitable for short-range transfers.

Bluetooth is a technology that connects and transfers information between electronic devices over a short-range radio frequency. It served for short-range data transfer in various smart clothing systems (Hung, Zhang, and Tai 2004; Naya et al. 2005). Hung, Zhang, and Tai (2004) investigated a combination of Bluetooth and wearable sensors to monitor in real time the vital signs of remote patients. Naya et al. (2005) proposed a Bluetooth-based indoor proximity detection method for location awareness in a nursing context. Proximity information exchanged between Bluetooth devices attached to people and medical apparatus helps estimate the location of nurses, patients, and medical equipment.

The MIT Media Lab developed a personal area network (PAN) that relies on the human body as a transmission medium in collaboration with IBM (Post et al. 1997). The network takes advantage of the natural salinity of the human body to transmit data; this salinity conducts electrical current well. This technology can transmit electronic data, e.g., business cards during a handshake. Security is a problem because touching a body equipped with a PAN could be like tapping into a telephone line, but the technique is convenient and cost-effective.

1.2.2.2   Long-Range Communication

Wireless connections are necessary for large-area communication systems. Among various communication systems, the Global System for Mobile communications (GSM) is suitable for small-sized data transfer such as voice transmission. The Third Generation (3G) wireless system can transfer files such as pictures and videos more quickly.

1.2.3 Data Management

Data management relates to memory, computation, and data processing. Electronic components are still used for those tasks since no textile material can perform them yet. Yet, miniaturization of the electronic components and development of flexible substrates has progressed so that the electronics can be incorporated into clothing in less obtrusive ways.

1.2.4 Energy Management

The major problem of wearable electronics is that conventional power supply is bulky, heavy, and rigid, with a short lifetime. Usually the biggest and heaviest part of a wearable device is the energy supply and storage, which also needs recharges at usual levels of power consumption. Power supply must be flexible and light enough to be incorporated into clothing without being a burden to the wearer. Supply should last long and recharge easily on the move, or use alternative energy sources that do not need recharging. The equipment should resist washing and wearing.

Several approaches were attempted to reduce the size and weight of electronic components for unobtrusive integration into clothing and to develop novel energy supplies from sunlight, heat, vibration, etc.

Bharatula, Zinniker, and Tröster (2005) developed a miniaturized hybrid micropower supply for wearable-pervasive sensors, using a lithium-ion button battery and a photovoltaic module, with power management circuitry. Power management techniques enable electronic devices to perform more efficiently with limited power, extending the battery-powered electronics' life.

Alternative energy sources include solar energy, body heat, and mechanical stimuli. Photovoltaic (PV) technology, especially thin film technology that enables the production of flexible structures, has gained much attention. In the solar-powered jacket from Scottevest Inc. (www.scottevest.com), thin film PV copper indium gallium diselenide (CIGS), a flexible photovoltaic material, was placed onto a thin stainless steel substrate, forming a detachable solar panel. Thin film PV materials such as amorphous silicon or CIGS can be layered onto a flexible polymeric film. Baps, Eber-Koyuncu, and Koyuncu (2002) adapted flexible solar cell technology to fiber form. A flexible tubular solar cell was assembled from stainless steel yarn as internal electrode, semiconductive nanosized TiO_2 powder, and organic dyes to sensitize the TiO_2 powder. The electrode, ceramic powder, and electrolyte were incorporated into a transparent polymer tube coated with conductive polymer on the inside.

Another energy supply method is thermoelectric conversion, transforming the temperature difference between the human body and environment into electricity. Infineon (Jung et al. 2003) miniaturized a silicon-based thermogenerator to harness and convert body heat into electricity; the generator was integrated into the fabric of the clothes for good thermal contact with the skin, leading to a large temperature difference at the interface.

As for mechanical motion, human movements can stimulate vibrational microgenerators, thus producing energy. For example, piezoelectric inserts in a shoe can generate electricity from heel strikes. In smart clothing, piezoelectric polymer materials are better candidates than non-textile piezoelectric ones in terms of flexibility and comfort.

1.2.5 INTEGRATED CIRCUITS

Integrated circuits are made out of semiconductor materials, with silicon as the most common for fabrication. However, the rigidity of silicon chips led to a search for flexible alternative materials; conductive polymeric materials are promising materials since they are flexible, lightweight, and robust (Tao 2005a).

There has been significant progress in smart clothing technology since the first generation of wearable computers. Thanks to advances in integration, bulky and rigid, portable machines attached to wearable systems were embedded into clothing. New materials, structures, and techniques led to commercialization. Biomonitoring smart clothing such as LifeShirt® is commercially available. Yet, a larger part still remains to develop smart clothing in which all components are made of full textile materials. Active research on textile-based integrated circuits and energy supplies is required. For real wearability, maintenance and durability must be addressed as well. Human aspects, another integral issue in smart clothing, are discussed in the following section.

1.3 HUMAN ASPECTS

Since smart clothing pursues the integration of clothing and electronic devices, it must provide functions that users want their clothing to exhibit, not the mere miniaturization of electronics. For smart clothing to effectively interact with users and environments, we must consider human aspects.

In smart clothing, human aspects derive from the integrated characteristics of clothing and electronic devices. As shown in Figure 1.4, the characteristics of electronic devices are usability, functionality, durability, and safety, while the characteristics of clothing include comfort, fashion, durability, and safety. Thus smart clothing should simultaneously provide the usability and functionality of electronic devices and the comfort and fashion of clothing, in addition to safety and durability, which are common to clothing and electronics.

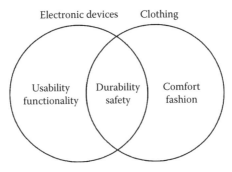

FIGURE 1.4   Human aspects in smart clothing.

1.3.1 Usability

Usability is a key factor for interfaces of smart clothes; they should provide easy input and output interfaces. As Nielson (1993) pointed out, usability has multiple components and is traditionally associated with learnability, efficiency, memorability, low error rate, and satisfaction.

- The system should be easy to learn so that users can rapidly get some work done.
- The system should be efficient to use, so that once a user has learned the system, productivity is high.
- The system should be easy to remember, so that a casual user can return to the system after a long period, without having to re-learn everything.
- The system should favor few errors by users. When and if errors occur, users should be able to recover from them easily. Furthermore, catastrophic errors must not occur.
- Using the system should be pleasant.

ISO 9241-11 (ISO 1988) suggests measuring usability based on effectiveness, efficiency, and satisfaction. Therefore, usability in smart clothing could be a function of the cognitive requirements associated with interactive matters.

Research on usability in smart clothing and wearable computing can be found in several papers. Gorlenko and Merrick (2003) discussed the usability of smart clothing in a general way. Besides, Chae, Hong, Cho et al. (2007) and Chae, Hong, Kim, et al. (2007) focused on the establishment of a usability test tool and on the evaluation of products using it.

Gorlenko and Merrick (2003) discussed the challenges of usability for mobile wireless computing. The most obvious challenge is dealing with mobility. The complexity and diversity of wearables presents serious challenges for the design process and for the design methodology. To ensure products are user centered, usability assessment should incorporate the user in the design process, reducing the production of expensive and time-consuming physical prototypes.

Chae, Hong, Cho et al. (2007) developed a usability evaluation tool consisting of 53 questionnaires on the usability of smart clothing. These questionnaires, designed based on observations and wearing tests, include five categories such as social acceptance, feeling of wearing, utility, easiness of maintenance, and safety. The tool was applied to evaluate an MP3 smart clothing product (Chae, Hong, Kim et al. 2007). Results indicated that improving the keypad interface is the most frequent request.

1.3.2 Functionality

Wearable technology should be applied to interfaces, communication, energy management, data management, and integrated circuits for integration in smart clothes. The functionality of each technology should be evaluated. Textile-based input interfaces and communication devices have been actively developed, leading to textile-based keypads, textile electrodes for interface technology, and textile-based signal transmission lines for wireless communication technology. Functionality of smart clothing has been considered in several studies, and we introduce hereafter three works regarding functionality tests of keypads and ECG electrodes.

To test the functionality of keypads, Cho et al. (2007a) compared subjective evaluations for, and measured operation forces of, two types of keypads: one with metal domes and one with rubber domes. Results showed that rubber dome keypads with 62 gf of operation force were preferred to metal dome keypads with 320 gf of operation force (Figure 1.5). Likewise, optimization of the feedback is critical for keypad construction.

Research was carried out on the performance of textile-based ECG electrodes to replace AgCl ECG electrodes commercially available for medical use. Jang et al. (2007) developed electrodes made of Cu-sputtered polyester fabric, and Cho, Jang, and Cho (2007b) constructed textile-based electrodes by embroidering stainless steel yarns on a cotton fabric.

To make Cu-sputtered fabrics for ECG electrodes, Cu-sputtering was performed on water-resistant nylon fabrics in a vacuum chamber by applying a high voltage across a low-pressure argon gas. To measure ECG, 1.5 cm × 3 cm textile electrodes were prepared to be placed on both wrists and on the right ankle, replacing conventional AgCl electrodes. Generally, the signal pattern of AgCl and Cu-sputtered textile electrodes was similar (Figure 1.6). However, Cu-sputtered electrodes showed a wider voltage range due to a wider contact area of the textile electrodes than the 1-cm-diameter AgCl electrodes.

The electrodes constructed by embroidering were attached to a man's sleeveless T-shirts for contact on the right and left parts of the chest and on the left abdomen. Then R-peak detection rates were obtained during wear tests (Figure 1.7). R-peak is the summit of the ECG signals and represents man's ventricular activity, and R-peak detection rates are the ratios of the numbers of R-peaks measured by the developed embroidered electrodes over those measured by the AgCl electrodes, and explain the degree of the ECG signal's accuracy. The embroidered electrodes had lower accuracy than Cu-sputtered textile electrodes because of the lower conductance of stainless steel yarn and of a much smaller contact area.

1.3.3 DURABILITY

Durability is important because clothing should withstand harsh conditions during laundering and everyday use. Electronic devices should be protected by a soft cushion or detached before laundering. The durability of electronics must be tested when they are integrated in textile form.

Cho et al. (2007a) and Park and Jayaraman (2001) discussed durability issues in developing smart clothing or electronic textiles. The Georgia Tech Wearable Motherboard-based shirt developed by Park and Jayaraman in 2001 to save the life of soldiers has a wear life of 120 combat days, which represents its ability to withstand repeated flexure, abrasion, and laundering.

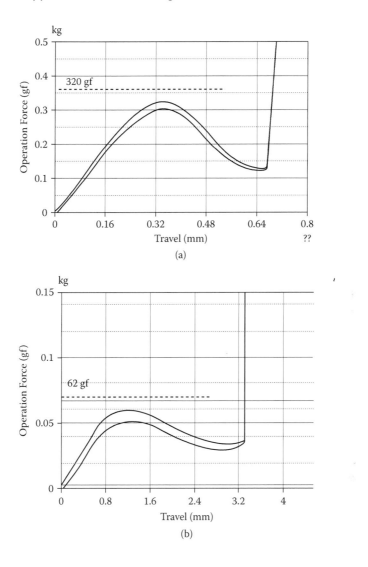

FIGURE 1.5 Operation forces of switches for textile-based keypads. (a) Metal dome switch. (b) Rubber dome switch.

To prove the durability of the fabric-based transmission lines, Cho et al. (2007b) developed textile-based transmission lines by applying polyurethane sealing on conductive fabrics. Polyurethane sealed metal-plated fabrics maintained excellent electrical durability even after 10 launderings. Moreover, when the earphone lines made of the laundered fabric were connected to an MP3 player, they worked successfully and no sound quality difference was subjectively perceived. Besides, the metal plated fabric with PU double sealing exhibited performance similar to conventional Cu earphone lines (Figure 1.8).

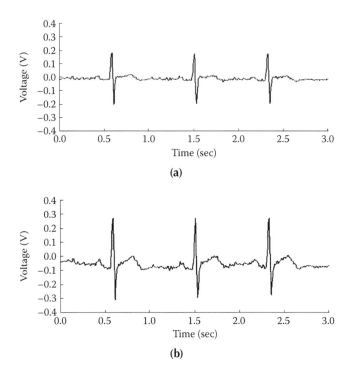

FIGURE 1.6 ECG signals of AgCl electrode and Cu-sputtered fabric. (a) AgCl electrode (b) Cu-sputtered electrode. (From "Exploring possibilities of ECG electrodes for bio-monitoring smartwear with Cu sputtered fabrics," by S. Jang, J. Cho, K. Jeong, and G. Cho, 2007, *Proceedings of HCI International 2007*, Fig. 6, p. 1136. Copyright 2007 by Springer. Reprinted with the kind permission of Springer Science + Business Media).

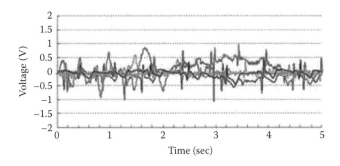

FIGURE 1.7 ECG signals of ECG shirt embroidered with stainless steel yarns.

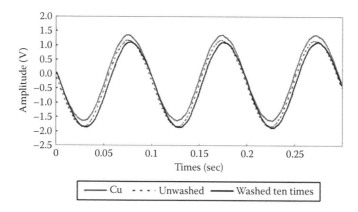

FIGURE 1.8 Comparison of output signals of textile-based cable. (From "Design and evaluation of textile-based signal transmission lines and keypads for smart wear," by J. Cho, J. Moon, M. Sung, K. Jeong, and G. Cho, 2007, *Proceedings of HCI International 2007*, Fig. 2, p. 1082. Copyright 2007 by Springer. Reprinted with the kind permission of Springer Science + Business Media).

1.3.4 SAFETY

Safety relates to being protected against physical, social, psychological, or other types of harm. The physical failure of smart clothing, such as overheating or electric shock, may be accrued due to functional error. Providing safe smart clothing requires consideration of physical forms, electromagnetic waves, electricity, etc.

To improve the safety of conductive narrow fabric for signal transmission, Yang et al. (2007) investigated whether adding Teflon coating on metal yarns used to construct conductive narrow fabric enhances safety. After applying external physical force (bending/abrasion) to the fabric, we verified that the narrow fabric utilizing Teflon-coated signal transmission lines maintained better electric insulation and prevented electrical interference (Figure 1.9).

Before mass-producing smart clothing, safety should be studied further from more diverse aspects; a standard index should accordingly be established for both manufacturers and consumers.

1.3.5 COMFORT

Comfort is freedom from discomfort and pain. According to Hatch (1993), comfort can be divided into thermophysiological comfort, sensorial or neurophysiological comfort, and body-movement comfort.

"Thermophysiological comfort" relates to the way in which clothing affects heat, moisture, and air transfers as well as the way in which the body interacts with clothing (Barfield et al. 2001). "Sensorial or neurophysiological comfort" relates to how consumers feel when clothing comes into contact with the skin. Finally, "body-movement comfort" relates to the ability of clothing to allow free movement, reduce burden, and to support the body.

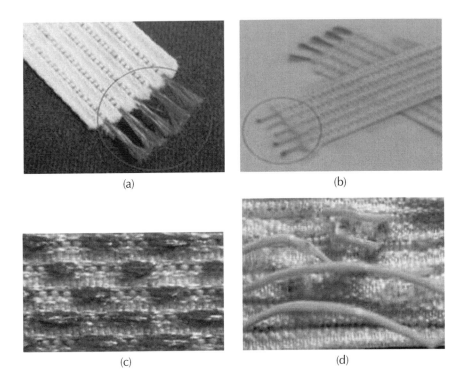

FIGURE 1.9   Conductive narrow fabrics for signal transmission lines before and after abrasion. (a) Without Teflon coating (before abrasion). (b) With Teflon coating (before abrasion). (c) Without Teflon coating (after abrasion). (d) With Teflon coating (after abrasion).

Comfort was assessed for wearable devices such as helmets or arm-worn devices (Robinette and Whitestone 1994; Stein et al. 1998; Whitestone 1993). However, assessments neglected various aspects.

Cho, Jang, and Cho (2007) developed ECG monitoring shirts varying in the spandex content (0%, 5%, and 7%) and conducted subjective evaluation through wear tests, using questionnaires about tightness, irritation, and ease of movement; the shirt with 5% spandex was preferred.

Knight and Baber carried out most research on the evaluation of smart clothing. They developed the comfort rating scales (CRSs) to assess wearable technologies (Knight et al. 2002; Knight and Barber 2005). The CRSs attempt a comprehensive assessment of wearer's comfort by measuring six dimensions: (1) emotion (concerns about appearance and relaxation); (2) attachment (comfort related to non-harmful physical sensation of the device on the body); (3) harm (physical sensation conveying pain); (4) perceived change (non-harmful indirect physical sensation making the wearer feel different overall, with perceptions such as being awkward or uncoordinated); (5) movement (awareness or modification to posture or movement due to direct impedance of inhibition by the device); and (6) anxiety (worries as to the safety of wearing the device and concerns as to whether it is used correctly or works properly).

The CRSs run on a 20-point scale and require users to rate their agreement from "low" to "high" to statements associated with each dimension. The CRSs were used to assess the comfort of numerous wearable technologies, in different situations (Bodine and Gemperle 2003; Knight et al. 2002; Knight and Baber 2005).

When compared to the conventional criteria for comfort (thermophysiological, sensorial, and body-movement comfort), CRSs deal with neuropsychological aspects as much as with physical and physiological aspects.

Gemperle and coworkers (1998) suggested design guidelines to communicate the considerations and principles necessary to design wearable products. The guidelines cover placement, form language, human movement, proxemics (human perception of space), size variation, attachment, containment, weight, accessibility, sensory interaction, temperature, aesthetics, and long-term use. The design guidelines help one to consider all creation issues for wearable devices.

Among the guidelines, proxemics provides a new viewpoint for the design of smart clothing. It refers to the understanding of layers of perception around the body, related to the fact that the brain perceives an aura around the body. It indicates the distance between an object and a user at which the object remains in the wearer's intimate space and is felt as a part of the body: 0 to 5 inches off the body (Gemperle et al. 1998).

1.3.6 FASHION

Since clothing is a fashion item, aesthetics should be combined with other human aspects for the balance of functional and aesthetic considerations. Designing smart clothing crosses the boundaries of specialists' knowledge. Some researchers in clothing and textile design have asserted the importance of multidisciplinary collaboration, and suggested development processes for smart clothing products (Ariyatum and Holland 2003; Baurley 2005; McCann, Hurford, and Martin 2005).

Fashion design and product design are established fields of their own, and it is difficult to communicate or adopt the others' work methods. Ariyatum and Holland (2003) asserted that a new product design model must be formulated based on a smart clothing context. The key issue they presented was that the conventional structure of new product design models fails to demonstrate the different work methods in electronics and in fashion. Thus, a new product design model is needed to enhance understanding about the work and communication with collaborators.

Baurley (2005) suggested a design methodology for interaction design in smart clothing; it consists of a conceptual framework, user studies, and design building. The framework relies on observations and research on how people use, interact with, and experience conventional clothes. User studies are based on user groups, and are fed back into the framework.

McCann, Hurford, and Martin (2005) proposed the "Critical Path" design tool to guide the design research and development process in the application of smart technologies. It was developed to support innovative decision making in the sourcing and selection of materials, technologies, and construction methods. The process includes identifying end-user needs, fiber and fabric development and textile assembly, and garment development. To balance appearance and function, designers require guidance in their selection and application of technical textiles, style, cutting, sewing, and finishing at every stage in the design research and development process.

1.4 APPLICATIONS

Smart clothing may serve in various fields since it offers functions for information, assistance, communication, aesthetics, etc. Some products are on the market, but generally developments are in a starting phase; their potential is enormous. However, no existing smart clothing fully integrates high technology and fashion design because the contributions from the electronics and fashion industries are unbalanced. This chapter illustrates the current situation of product development and explains the technology involved in integrating electronics into fashion.

1.4.1 BODY MONITORING

Truly instrumented garments that monitor physiological, neurological, and body kinematic parameters are crucial for healthcare and health provision (De Rossi et al. 2000). As Lymberis and Olsson stated in 2003, they can provide physicians with data to detect and manage health risks, diagnose at early stages, recommend treatments, and make professional decisions based on objective information (Lymberis and Olsson 2003).

Healthcare is a key market for the textile industry: in 2000, over 1.5 million tons of medical and hygienic textile materials, worth $5.4 billion, were consumed worldwide. This figure should increase in volume by 4.5% per annum, reaching 2.4 million tons by 2010, with a market value of $8.2 billion (David Rigby Associates 2002). We focus hereafter on the analysis of applications potentially beneficial to users and on the enabling technology required for this vision to happen.

1.4.1.1 Body Signals

Medical monitors benefited from technological advances in wireless communication, processing, and power sources, allowing miniaturization and prolonged operating times, as well as global integration into telemedical systems. Medical monitors provide patients with real-time feedback about medical conditions related to respiration, ECG, GSR (galvanic skin response), skin pH, temperature, and blood oximetry while going about their normal daily activities. They also inform athletes during training and healthy users about their physiological state (Raskovic, Martin, and Jovanov 2003).

The "SmartShirt" wearable motherboard developed by Park and Jayaraman in 2001 can support persons affected by known disorders: it permits continuous monitoring of their physical condition by medical personnel in a non-invasive manner. The goal of the system is to minimize casualties for first responders by providing knowledge of their health status to on-scene commanders in diverse emergencies. The SmartShirt monitors various vital signs such as heart rate, ECG, respiration, and blood pressure. The initial system included three-lead ECG electrodes, a heart rate monitor, and a respiration monitor; it uses electro-optical fibers embedded in the fabric to collect biomedical information sent to a transmitter where it is stored or wirelessly sent to a doctor, coach, or personal server using Bluetooth, RF, wLAN, or cellular signals. This item contains basic smart clothing components, but uses only a limited textile technology, for communication.

VivoMetrics' flagship product, LifeShirt®, is an ambulatory vest that monitors respiration with thoracic and abdominal inductive plethysmography bands sewn into a Lycra® vest. The shirt's input system is functionalized with carbon-loaded rubber (CLR) piezoresistive fabric sensors, used to monitor respiration trace (Grossman 2003). LifeShirt® also includes an ECG sensor normally used for medical purposes. The gathered data are stored in a recorder. This management system is incorporated into a customized Handspring worn on the patient's belt or carried in a pocket. The patient will upload the data via the Internet to VivoMetrics' secure data center. Then, it is reviewed by technicians and physicians. LifeShirt® collects, analyzes, and reports on the subject's pulmonary, cardiac, and posture data (Solaz et al. 2006).

Taccini et al. (2004) presented the WEALTHY system, which integrates function modules such as sensing, conditioning, pre-processing, data transmission, and remote monitoring. The WEALTHY system also contains a prototype of respiration sensing device that derives the respiration of the wearer from impedance. The main respiration and movement activity come from piezoresistive sensors used as input interface, sampled at 16 Hz. Their signals are transmitted without local processing. The coated Lycra® fabric detects respiration signals. Strain fabric sensors based on piezoresistive yarns and the fabric electrodes realized with metal-based yarns enable the realization of wearable and wireless instrumented garments capable of recording physiological signals. The WEALTHY system is for routine activities and should replace a classical garment without discomfort to the user. The main innovation is functionalized materials in the form of fibers and yarns knitted or woven into a multifunctional sensing fabric (Taccini et al. 2004).

Lind et al. (1997) designed Sensate Liner (SL) to efficiently monitor the medical condition of personnel at a manageable cost with an instrumented uniform. The textile consists of a mesh of electrically and optically conductive fibers integrated into a normal woven or knitted structure of fibers and yarns selected for comfort and durability. SL consists of a form fitting two-piece jumpsuit containing and connecting sensors and devices to a pack containing a processor and a transmitter. The structure contains regularly spaced yarns acting as sensing elements, and precisely positioned yarns carrying signals from the sensors to the Personal Status Monitor. The fabric substrates are suitable for the incorporation of optical sensors, reflective or camouflage materials, communications networking, or fiber optic cable interfaces.

Ottenbacher et al. (2004) developed a shirt with ECG electrodes as input interface, copper wires, and Bluetooth for communication. A very thin and flexible edging by the TITV was employed; it consists of eight isolated copper wires, protection mesh, and a border for sewing. Two wires convey the signal while the others provide shielding. The electronic hardware is removable, to realize a washable system. Pushbuttons are used to provide an easy way to connect the hardware to the shirt electrically and mechanically. Two pushbuttons are used for each signal, avoiding contact problems.

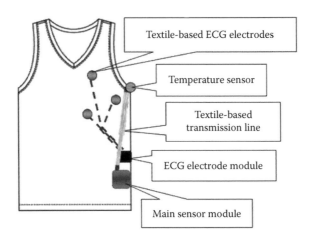

FIGURE 1.10   Diagram of a jogging wear.

Using integrated textile electrodes, the electrodes remain in place, and skin does not become irritated even after long use.

As mentioned earlier, Van Langenhove and Hertleer (2004) developed the Respibelt to measure respiration. It is a fabric sensor made of a stainless steel yarn, knitted in a Lycra® belt, providing an adjustable stretch. Using piezoelectric sensors, Respibelt can sense the wearer's respiration. By placing the Respibelt around the abdomen or thorax, circumference and length changes of the Respibelt, caused by breathing, result both in an inductance and resistance variation.

The Smart Wear Research Center at Yonsei University developed jogging wear to monitor personal health conditions while jogging by sensing heart rates and temperature. As shown in Figure 1.10, it comprises three textile-based ECG sensors, a temperature sensor, textile-based transmission lines, ECG electrode module, and the main sensor module. The ECG sensors are made of metal-plated conductive fabrics, replacing the conventional adhesive type of electrodes. Transmission lines are textile based, which make the garment washable.

1.4.1.2   Body Movements

Motions, motion patterns, gestures, and postures are basic elements characterizing human activity. Tracking body motions, gestures, and positions provides useful information to classify activities, to remove noise from biosignals, and to interpret physiological status (Tröster 2004). Accelerometers, gyroscopes, magnetometers, piezoelectric sensors, and GPS (global positioning system) are often combined to detect motion.

Wearable computers or smart clothing were used to study measurement methods for human motions in various fields. Motion capture systems work but cause inconveniences to users and require many devices such as cameras and image analyzing systems. Thus, textile-based motion sensors are an attractive alternative (Sung et al. 2007).

Sung et al. (2004) presented a wearable real-time shiver monitor based on the MIThril Live Net system, a flexible, distributed, mobile platform usable for various proactive healthcare applications. In this exploratory study, they demonstrated that shivering can be accurately determined from continuous accelerometer sensing. They attempted to develop a real-time wearable monitor to accurately classify shivering motions through simple accelerometer sensing and statistical machine learning techniques. Motion sensing is accomplished via two embedded microcontroller-based sensors based on the Analog Devices ADXL202 accelerometer part. As moderate cold exposure occurs, shivering becomes intense, uncontrollable. This exploratory research anticipates the emergence of real-time health monitoring systems capable of classifying the cold exposure of soldiers in harsh cold environments with non-invasive sensing and minimal embedded computational resources.

Motoi et al. (2006) developed a wearable system for monitoring walking speed, and static and dynamic posture, providing a quantitative assessment of a patient's motion during rehabilitation programs. In the dynamic posture, the angle change is obtained by integrating the gyro-sensor signal. The movements of four markers attached on the subjects are simultaneously recorded at 30 frames per second with a charge-coupled device (CCD) camera. Posture changes of subjects are also recorded using a digital video camera. Despite the precise evaluation, this system still needs to be flexible when applying the electronic textile technology.

Inductive fiber-meshed transducers (FMTs) developed by Wijesiriwardana (2006) can be integrated into sleeves, tights, and the calf of leggings of garments for angular measurements. Knitting technology is used for its construction with metals. A single coil based on variation in self-inductance or electromagnetic induction is used to measure angular displacement. The knitted input interface provides more body movement comfort than any other material.

Gibbs and Asada created wearable conductive fiber sensors to continuously measure joint movements in 2004. To measure these movements, an array of 11 conductive yarns across the knee joint is used. Each yarn is separated by 5 mm, and each has an unstretched length of 55 cm. The threads are silver plated nylon 66 yarns produced by embedding tiny particles of electrically conductive carbon powder into the surface of a nylon fiber.

Michahelles and Schiele (2005) developed a sensing and monitoring system for professional skiers. Based on wearable sensors and video recording, it can reveal important features of the athlete's motions, helping trainers identify the skier's strengths and weaknesses. They placed three force-sensing sensors in the shape of a triangle on the foot to obtain a center of pressure. When the skier's foot applies pressure, semi-conductive ink in the sensor shunts the electrodes, measured as a resistance value. The system can provide autonomous teaching of snowboarding, skateboarding, and cycling.

Taelman et al. (2006) suggested contactless electromyogram (EMG) sensors to continuously monitor muscle activity, to prevent musculoskeletal disorders. The sensors are developed for a wearable textile providing measures unobtrusively; they do not require skin contact because they detect an electric displacement current by means of a capacitive coupling to the body. The sensors are integrated by embroidery with conductive and insulated threads. The lack of need for skin contact gives the opportunity to wear the vest above other clothing, and enhances comfort.

Rocha and Correia (2006) designed a wearable sensor network to monitor body kinematics such as posture, gesture, heart rate, respiratory rate, and temperature of a patient during treatment. The sensing modules, composed of three-axis accelerometers, three-axis magnetometers, and interface electronics, are integrated in a swimming suit and connected to a microcontroller by a serial interface. The suit integrates both floats and electronic components.

The main characteristic of the application is the wet environment, which adds unique development constraints. Customized sensing microsystems inserted in waterproof pockets can be integrated in the textiles. The body kinematics monitor microsystem is a sensor network composed of five modules. Both the gravitational force and the earth's magnetic field are used to detect the posture of the main body articulations.

Dunne et al. (2005) developed garments that gather body context information. The key component is a pressure-sensitive polypyrrole-coated foam. Instead of monitoring changes in limb position, body length, or circumferences, information is gathered from the pre-existing dynamic physical forces that operate between the wearer's body and a garment during physiological functions or movements. The test configuration uses foam to monitor breathing, shoulder movements, neck movements, and constant pressure on the shoulder blades. Made from standard polyurethane foam with a conductive polypyrrole coating, the sensor retains all of the physical properties of regular polyurethane foam, and is washable (Brady et al. 2005). Importantly, this technology can be readily embedded into a normal garment, retaining the structural and tactile properties of a textile structure. Based on preliminary data, the polypyrrole-coated conductive foam is promising as a basic sensing technology for detecting body movements, physiological functions, and body states from body-garment interactions.

Work-related upper limb musculoskeletal disorders among computer users can be prevented by posture modification (Gerr et al. 2005), or ergonomic interventions like specific hardware (e.g., adjustable chairs) or workstation modifications (Lindegård et al. 2005). However, few reliable, objective, and accurate methods allow continuous monitoring of posture at work to evaluate the success of these interventions.

Dunne et al. (2006) used a wearable POF sensor to monitor seated spinal posture. A fiber optic bend sensor consists of a light source, a light sensor, and a length of plastic optic fiber (POF). The POF is abraded along one side, allowing light to escape. The amount of light sensed depends on the bend of the fiber; the sensor response is reliable, accurate, repeatable, and drift-less. The POF sensor is inexpensive and well suited to wearable applications, because of its small size, flexibility, and easily customized length.

1.4.2 ENTERTAINMENT

Several sportswear companies successfully sell products by adding values related to information and entertainment. Now, smart clothing appealing to the entertainment world is getting attention by interacting with emotional effect.

Post et al. (2000) built Musical Jackets with a touch-sensitive MIDI keyboard embroidered directly into the fabric using conductive thread. The thread, embroidered into a standard 4 × 3 character keypad below the right shoulder, contains stainless steel filaments, making it conductive. The capacitive loading of the body is detected when the thread is touched; the keypad is polyphonic, thus several keys can be hit simultaneously. Sound is generated by a single-chip General MIDI wavetable synthesizer, and sequences are generated by a microcontroller. The jacket is entirely battery operated, with powered speakers in the pockets.

The Smart Wear Research Center presented an MP3 player dress containing textile-based transmission lines and textile-based switches. Switches are freely placed on the front bodice, representing different symbols for MP3 player operation.

The Smart Wear Research Center also presented a photonic dress whose fabric surface illuminates when combined with an LED light source. The basic type of photonic dress relies on POF fabrication and on the etching technique for the POF surface. To evenly illuminate the POF fabric, the etching should be uniform. Besides, the amount of etching should be controlled to ensure fabric durability.

The color-responding photonic dress changes its color according to the environment. When a color is sensed by the sensor, it is characterized by a specific value, then matched to three color LEDs, which emit light through the POF fabric.

The sound-responding photonic jacket (Figure 1.11) changes its color according to environmental sound. A microcondenser senses low frequency sound, and corresponding frequencies (e.g., drum sound) change resistance and capacitor. By connecting this sound-sensing module to an LED, the POF fabrics illuminate like an equalizer. The components of the sound-responding photonic jacket are POF fabrics with an LED, an LED control module, and rechargeable batteries.

Vilkas and Kukkia, designed by Berzowska and Marcelo (2005), are expressive and behavioral kinetic sculptures. Vilkas is a dress with a kinetic hemline on the right side that rises over a 30-second interval to reveal the knee and lower thigh. It is constructed of heavy hand-made felt that contracts through the use of hand-stitched Nitinol wires. Nitinol, a shape memory alloy (SMA) made of nickel and titanium, can indefinitely remember its geometry if once treated to acquire a specific shape. Using this characteristic, the Nitinol can create a wrinkling effect.

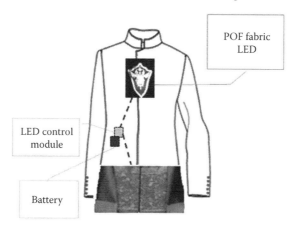

FIGURE 1.11   Sound-responding photonic jacket.

The Kukkia dress is decorated with three animated flowers that open and close over a 15-second interval. When heated, wire shrinks and pulls the petals together, closing the flower. As it cools down, the rigidity of the felt counteracts the shape of the wire, opening the flower. This dress is operated by a microcontroller triggering drivers that send power to the Nitinol. Also, it uses small rechargeable lithium polymer cells that can power the jacket for 2 hours. The shape memory embedded fabric is used as output interface.

Touch is a powerful conduit for emotional connectedness. Haptic research includes the design of interactions employing devices through which virtual physical models can be felt, just as we display to our visual sense with graphical displays. Using haptic communication, separated individuals interact with one another through a pair of haptic displays, themselves connected via a computer running coupled virtual physical models (Smith 2007).

The Sensor Sleeve designed by Randell et al. (2005) is a cloth that enables the remote exchange of emotional messages between intimate people, by conveying a sense of touch and presence. The sleeve detects embrace and stroke actions on the arm. A smart textile system comprising gesture and touch sensors, a microcontroller, and Bluetooth devices is integrated into the sleeves.

Cutecircuit's Hug Shirt (http://www.cutecircuit.com) is a shirt that enables people to hug over a distance. Detachable pads contain sensors that sense touch pressure and heart beat. The actuators reproduce the sensation of touch and warmth. When touching the red areas of the shirt, the mobile phone receives sensor data via Bluetooth and delivers them to the partner. The Hug Shirt is built using textile pad type of interfaces, wireless communication, integrated circuit, and rechargeable batteries. The Hug Shirt may be particularly useful to soldiers to contact more intimately their loved ones far away. Nominated as a 2006 innovation of the year by *TIME* magazine, the Hug Shirt is being prepared for the market.

Dunne (2004) designed massage shirts to use vibrotactility for communication and information display. During tests of the vibrotactile shoulder pad display, a common subject remark to the tactile stimulus was that it felt "relaxing," "comforting," or "soothing." Thus, a massage function based on the same vibrating motors may be attractive. To achieve the massage, each shirt contained six to eight flat vibrating motors powered by one 9-volt battery.

Communication-Wear designed by Sharon Baurley's research team is a clothing concept that augments the mobile phone by enabling expressive messages to be exchanged remotely with a sense of touch and presence (Baurley et al. 2007). When a hug or embrace gesture is sent, the heat pads in the back of the jacket heat up symbolizing the warming sensation felt when touched by another person. Communication of touch messages takes place between garments via Bluetooth. Physiological arousal, evaluated from GSR sensors, is relayed to the partner via light emitted by the fiber optic section.

This garment contains a circuit board where a PIC microcontroller for processing is mounted, and a 7.2-volt re-chargeable battery. This Communication-Wear also contains input interfaces like touch sensing method or textile-based GSR sensor, output interface (heat-emitting fabric, light emitting fabric), communication interface such as Bluetooth, integrated circuit, and battery.

Affective computing grows out of wearable computing as suggested by Picard and Healey (1997), motivating the creation of smart clothing that can recognize physical and psychological patterns and then translate them into emotions. Emotions notably influence the autonomic nervous system activity and thus the heart rate and skin conductivity (Baurley 2005). This is valuable to engineering smart clothing equipped with a sensing function for wearer's physiological signals, to analyze the collected data, and to react appropriately towards wearers.

The Scentsory Design® (http://www.smartsecondskin.com) project exploits scent to improve mental and physical well-being. The designs speak for the wearer through smells by interpreting emotions, enabling the wearer to express his or her feelings through the delivery of color and scent emitted from clothing. Thus, the item produces a very personal scent, delivering sensations on demand. Such smart clothes can alleviate mental and physical health problems through the delivery of odorant-beneficial chemicals in controlled ways responding to personal needs.

Marrin (1999) developed the Conductor's Jacket, a physiological sensing systems robust to motion artifacts. This wearable system is used to detect the relationships of the conductor's musical gestures with the patterns of muscle tension and breathing. Seven EMG and a respiration sensors provide input. The EMG sensors are fixed with custom-fit elastics sewn into the shirt, so that they remain snug without strong adhesives, and yet do not move when arms move. This system was designed to measure how professional and student conductors naturally communicate expressive information to an orchestra. After analyzing real conducting data from six subjects, Marrin (1999) found 30 significant expressive features related to muscle tension changes.

1.4.3 INFORMATION

Portable devices have spread tremendously in recent years. These products are designed to be carried near the body. The demand for smart clothing will grow as applications for networked computer-based devices on the body multiply and diversify. Potential applications go beyond communication, emails, and organization of personal data.

These devices have their own input and output devices. To provide clothes with information functions, input and output interfaces should be designed for better efficiency.

The Communication Jacket developed by a Fraunhofer research team integrates a mobile phone into clothing. The I/O modules are connected by embroidered lines and metal snaps that allow easy removal of the display module for washing. Although the input device is made of textile, the output device is packaged in a hard case, preventing complete integration.

Randell developed a CyberJacket in 2001 with the Bristol Wearable Computing project team that provides a platform for developing and testing wearable applications. It includes a network computer and numerous context-sensing devices (a GPS receiver, ultrasonic indoor location sensors, electronic compass, accelerometers). The interface includes speech recognition and audio playback, and displays can be handheld, head mounted, or worn on the sleeve. When the CyberJacket detects that the user is approaching a certain place, it plays a short audio message through an earpiece. If the user stops, relevant information appears on the sleeve. The devices in the CyberJacket are still in the early stage of integration into clothing.

1.5 REAPPRAISAL OF SMART CLOTHING

We reviewed technology developments and applications of smart clothing. Based on the current status of smart clothing, this chapter suggests the direction to develop smart clothing and future work.

1.5.1 FUTURE DEVELOPMENTS OF SMART CLOTHING

With the growing interest in smart clothing from the industry as well as academia, we anticipate that the area of smart clothing will continue to expand. Growth is expected to occur in applications requiring dedicated functions and in everyday clothing in which the wearer's emotions could be recognized and expressed.

Figure 1.12 summarizes the components of smart clothing technology, the services that smart clothing can provide, and examples of the applications. Individual components of a smart clothing system, i.e., interface, communication, data management, energy management, and integrated circuits, are combined and work together to form services such as information, communication, assistance, aesthetic, affective, etc. An example of information application is a jacket with a GPS, providing positioning and location data. Communication applications include a jacket in which a mobile phone is integrated. An example of affective smart clothing is the Sensor Sleeve, which detects affective gestures and enables emotional messages to be exchanged remotely between people.

As illustrated in Figure 1.12, smart clothing is likely to expand from a function-oriented system to a system that focuses not only on the function but also affective states of the wearer. We are living in an era where human beings search for social interaction and relationships and look for ways to communicate with others and

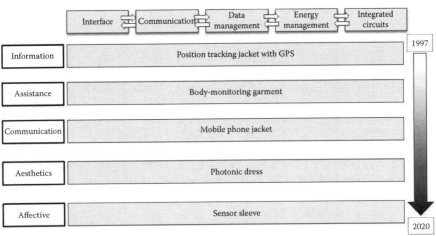

FIGURE 1.12 Smart clothing technology, service, and applications.

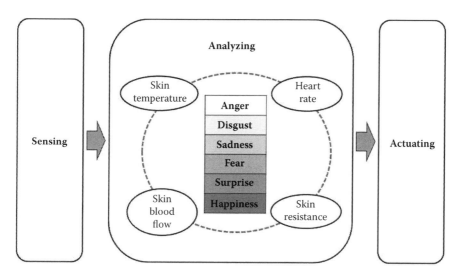

FIGURE 1.13 Sensory information monitoring mechanism.

express our emotions. In the future, smart clothing may sense the wearer's feelings and respond to emotions by changing its shape, color, scent, and so forth.

Figure 1.13 shows an example of how affective smart clothing works. It illustrates a sensory information monitoring mechanism where the wearer's emotions are read and interpreted. The wearer's physiological signals such as heart rate or skin temperature are sensed, the collected data are analyzed, and appropriate responses are provided through actuators. Future smart clothing systems would detect emotional information based on the wearer's physical state and respond to the wearer's emotions.

The already marketed smart clothing products discussed in the previous section focus mostly on embedding a single service or function into clothing. Research on smart clothing systems should expand to integrate various services or functions into the system to produce multi-functional smart clothing systems. In order to achieve smart clothing that is genuinely intelligent and interactive with human beings, novel materials and technologies are essential.

1.5.2 FUTURE WORKS

In smart technology for textiles and clothing, continued development is needed to integrate technology that combines electronics into each component of smart clothing. We expect to proceed from block-based technology to embedded technology, and accordingly to move technology from textile converting to fiber level. Therefore, novel fiber materials such as ICPs and new technologies such as nanotechnology and electrospinning will play an important role in the next-generation material/technology.

To implement the functions of each component in smart clothing, interconnection technology for bridging is also required. Conductive textiles in each component are currently at different technology levels, for example, the technological level in displays exceeds that of batteries. Thus, we must standardize specifications for efficient interconnections between components, considering interchangeability, durability, and usability.

Smart clothing is a fashion item that needs to satisfy the users in emotion as much as in function. Therefore, user-oriented technology development that reflects consumers' latent needs is essential, in addition to integration and interconnection technologies.

ACKNOWLEDGMENTS

This work was financially supported by grants (10016447 and 10016575) from the Ministry of Knowledge Economy, Republic of Korea, and supported by the Smart Wear Research Center.

REFERENCES

Ariyatum, B., and R. Holland. 2003. A strategic approach to new product development in smart clothing. *Proceedings of the 6th Asian Design Conference.* Tsukuba.

Baber, C. 2001. Wearable computers: A human factors review. *International Journal of Human-Computer Interaction* 13 (2): 23–145.

Baps, B., M. Eber-Koyuncu, and M. Koyuncu. 2002. Ceramic based solar cells in fiber form. *Key Engineering Materials* 206–13, 937–40.

Barfield, W., S. Mann, K. Baird, F. Gemperle, C. Kasabach, J. Stivoric, M. Bauer, R. Martin, and G. Cho. 2001. Computational clothing and accessories. In W. Barfield and T. Caudell (Eds.), *Fundamentals of wearable computers and augmented reality,* 471–509. Lawrence Erlbaum Associates, Inc.

Baurley, S. 2005. Interaction design in smart textiles clothing and applications. In T. Xiaoming (Ed.), *Wearable electronics and photonics,* 223–243. Woodhead Publishing Ltd. and CRC Press LLC.

Baurley, S., P. Brock, E. Geelhoed, and A. Moore. 2007. Communication-wear. *Proceedings of Ubicomp 2007 Adjunct—Transitive Materials: Towards an Integrated Approach to Material Technology,* a workshop of the 9th International Conference on Ubiquitous Computing (Ubicomp 2007), Innsbruck.

Berzowska, J., and C. Marcelo. 2005. Kukkia and vilkas: Electronic garments. *Proceedings of the 9th IEEE International Symposium on Wearable Computer* (ISWC 2005), Osaka.

Bharatula, N. B., R. Zinniker, and G. Tröster. 2005. Hybrid micropower supply for wearable-pervasive sensor nodes, In *Proceedings of the 9th IEEE International Symposium on Wearable Computers,* 196–7. Osaka, Japan: IEEE Computer Society.

Bodine, K., and F. Gemperle. 2003. Effects of functionality on perceived comfort of wearables. *Proceedings of the 7th IEEE International Symposium on Wearable Computers.* White Plains, NY.

Brady, S., L. E. Dunne, R. Tynan, D. Diamond, B. Smyth, and G. M. P. O'Hare. 2005. Garment-based monitoring of respiration rate using a foam pressure sensor, In *Proceedings of the 9th IEEE International Symposium on Wearable Computers,* 214–15. Osaka, Japan: IEEE Computer Society.

Catrysse, M., R. Puers, C. Hertleer, L. Van Lagenhove, H. van Egmond, and D. Matthys. 2004. Towards the integration of textile sensors in a wireless monitoring suit. *Sensors and Actuators A* 114:302–11.

Chae, H., J. Hong, H. Cho, K. Han, and J. Lee. 2007. An investigation of usability evaluation for smart clothing. *Proceedings of HCI International 2007,* 1053–60.

Chae, H., J. Hong, J. Kim, J. Kim, K. Han, and J. Lee. 2007. Usability evaluation and development of design prototyping for MP3 smart clothing product. *Korean Society for Emotion and Sensibility* 10 (3): 331–42.

Chan Vili, Y. Y. F. 2007. Investigating smart textiles based on shape memory materials. *Textile Research Journal* 77 (5): 290–300.

Cho, J., J. Moon, M. Sung, K. Jeong, and G. Cho. 2007a. Design and evaluation of textile-based signal transmission lines and keypads for smart wear. *Proceedings of HCI International 2007*, 1078–85.

Cho, J., S. Jang, and G. Cho. 2007b. Effects of fabric elasticity on performance of textile-based ECG-monitoring smart wear. *Proceedings of 2007 Spring Conference of Korean Society for Emotion and Sensibility,* 39–41.

David Rigby Associates. 2002. Technical textiles and industrial nonwovens: World market forecast to 2010. Manchester, U.K.: Author.

De Rossi, D., A. Della Santa, and A. Mazzoldi. 1999. Dressware: Wearable hardware. *Materials Science and Engineering C* 7:31–35.

De Rossi, D., F. Lorussi, A. Mazzoldi, P. Orsini, and E. P. Scilingo. 2000. Monitoring body kinematics and gesture through sensing fabrics. *Proceedings of the 1st Annual International IEEE-EMBS Special Topic Conference on Microtechnologies in Medicine and Biology*, 587–92. Lyon, France.

Dhawan, A., A. M. Seyam, T. K. Ghosh, and J. F. Muth. 2004. Woven fabric-based electrical circuits. Part I: Evaluating interconnect methods. *Textile Research Journal* 74 (10): 913–19.

Dunne, L. E. 2004. The design of wearable technology: Addressing the human-device interface through functional apparel design. Master's Thesis, Cornell University. Ithaca, NY, USA.

Dunne, L. E., S. Brady, B. Smyth, and D. Diamond. 2005. Minimally invasive gathering of body context information from garment interactions. *Proceedings of the 3rd UK-UbiNet Workshop*.

Dunne, L. E., P. Walsh, B. Smyth, and B. Caulfield. 2006. Design and evaluation of a wearable optical sensor for monitoring seated spinal posture. *Proceedings of 10th ISWC 2006*, 65–68.

El-Sherif, M. A., J. Yuan, and A. MacDiarmid. 2000. Fiber optic sensors and smart fabrics. *Journal of Intelligent Material Systems and Structures* 11:407–14.

Farringdon, J., A. J. Moore, N. Tilbury, J. Church, and P. D. Biemond. 1999. Wearable sensor badge and sensor jacket for context awareness. In *Digest of Papers of the 3rd International Symposium on Wearable Computers,* 107–13. Los Alamitos, CA: IEEE Computer Society.

Gemperle, F. C. Kasabach, J. Stivoric, M. Bauer, and R. Martin. 1998. Design for wearability. *Proceedings of the 2nd International Symposium on Wearable Computers*, 116–22. Los Alamitos, CA.

Gerr, F., M. Marcus, C. Monteilh, L. Hannan, D. Oritz, and D. Kleinbaum. 2005. A randomized controlled trial of postural interventions for prevention of musculoskeletal symptoms among computer users. *Occupational and Environmental Medicine* 62:478–87.

Gibbs, P., and H. H. Asada. 2004. Wearable conductive fiber sensor arrays for measuring multi-axis joint motion. *Proceedings of the 26th Annual International Conference of the IEEE EMBS,* 4755–58.

Gorlenko, L., and R. Merrick. 2003. No wires attached: Usability challenges in the connected mobile world. *IBM Systems Journal* 42 (4): 639–51.

Gorlick, M. M. 1999. Electric suspender: A fabric power bus and data network for wearable digital devices. In *Digest of Papers of the 3rd International Symposium on Wearable Computers*, 114–21. Los Alamitos, CA: IEEE Computer Society.

Grossman, P. 2003. The LifeShirts: A multi-function ambulatory system that monitors health, disease, and medical intervention in the real world. *Proceedings of the International Workshop—New Generation of Wearable Systems for eHealth,* 73–80.

Hatch, K. L. 1993. *Textile science.* Minneapolis: West Publishing.

Hung, K., Y. T. Zhang, and B. Tai. 2004. Wearable medical devices for tele-home healthcare. In *Proceedings of the 26th Annual International Conference of the IEEE EMBS*, 5384–87. San Francisco, CA: IEEE Computer Society.

International Standard ISO 9241–11:1988(E) Ergonomic requirements for office work with visual display terminals (VDTs)—Part 11: Guidance on usability.

Jang, S., J. Cho, K. Jeong, and G. Cho. 2007. Exploring possibilities of ECG electrodes for bio-monitoring smartwear with Cu sputtered fabrics. In *Proceedings of HCI International 2007,* 1130–37.

Jung, S., C. Lauterbach, M. Strasser, and W. Weber. 2003. Enabling technologies for disappearing electronics in smart textiles. In *Proceedings of the 2003 IEEE International Solid-State Circuits Conference*, 386–7. IEEE International.

Kirstein, T., D. Cottet, J. Grzyb, and G. Tröster. 2005. Wearable computing systems—Electronic textiles. In T. Xiaoming (Ed.), *Wearable electronics and photonics*, 177–97. Woodhead Publishing Ltd. and CRC Press LLC.

Knight, J. F., and C. Baber. 2005. A tool to assess the comfort of wearable computers. *Human Factors* 47 (1): 77–91.

Knight, J. F., C. Baber, A. Schwirtz, and H. W. Bristow. 2002. The comfort assessment of wearable computers. *Proceedings of the 6th International Symposium of Wearable Computers*, 65–72. Seattle.

Koncar, V., E. Deflin, and A. Weill. 2005. Communication apparel and optical fibre fabric display. In T. Xiaoming (Ed.), *Wearable electronics and photonics,* 155–76. Woodhead Publishing Ltd. and CRC Press LLC.

Lam Po Tang, S., and G. K. Stylios. 2006. An overview of smart technologies for clothing design and engineering. *International Journal of Clothing Science and Technology* 18 (2): 108–208.

Lane, R., and B. Craig. 2003. Materials that sense and respond: An introduction to smart materials. *The AMPTIAC Quarterly* 7 (2): 9–14.

Lind, E. J., S. Jayaraman, S. Park, R. Rajamanickam, R. Eisler, G. Burghart, and T. McKee. 1997. A senate liner for personal monitoring applications, In *Digest of Papers of the 1st International Symposium on Wearable Computers,* 98–105. Los Alamitos, CA: IEEE Computer Society.

Lindegård, A., C. Karlberga, E. W. Tornqvist, A. Toomingas, and M. Hagberg. 2005. Concordance between VDU-users' ratings of comfort and perceived exertion with experts' observations of workplace layout and working postures. *Applied Ergonomics* 36 (3): 319–25.

Loriga, G., N. Taccini, D. De Rossi, and R. Paradiso. 2005. Textile sensing interfaces for cardiopulmonary signs monitoring. *Proceedings of the 2005 IEEE Engineering in Medicine and Biology 27th Annual Conference*, 7349–52.

Lymberis, A., and S. Olsson. 2003. Intelligent biomedical clothing for personal health and disease management: State of the art and future vision. *Telemedicine Journal and e-Health* 9 (4): 379–86.

Mann, S. 1996. Smart clothing: The shift to wearable computing. *Communications of the ACM* 39 (8): 23–24.

Marrin, T. 1999. Inside the Conductor's Jacket: Analysis, interpretation, and musical synthesis of expressive gesture, Ph.D. Thesis, MIT, Cambridge, MA, USA.

McCann, J., R. Hurford, and A. Martin. 2005. A design process for the development of innovative smart clothing that addresses end-user needs from technical, functional, aesthetic and cultural view points. *Proceedings of IEEE International Symposium on Wearable Computers* (ISWC 2005), 70–77.

Michahelles, F., and B. Schiele. 2005. Sensing and monitoring professional skiers. *Proceedings of the IEEE Pervasive Computing, 2005,* 40–46.

Motoi, K., K. Ikeda, Y. Kuwae, M. Ogata, K. Fujita, D. Oikawa, T. Yuji, Y. Higashi, T. Fujimoto, M. Nogawa, S. Tanaka, and K. Yamakoshi. 2006. Development of a wearable sensor system for monitoring static and dynamic posture together with walking speed for use in rehabilitation. *Proceedings of World Congress on Medical Physics and Biomedical Engineering*, CD-ROM.

Naya, F., H. Noma, R. Ohmura, and K. Kogure. 2005. Bluetooth-based indoor proximity sensing for nursing context awareness. In *Proceedings of the 9th IEEE International Symposium on Wearable Computers*, 212–13. Osaka, Japan: IEEE Computer Society.

Nielson, J. 1993. *Usability engineering.* San Francisco: Morgan Kaufmann Publishers, Inc.

Ottenbacher, J., S. Romer, C. Kunze, U. Grosmann, and W. Stork. 2004. Integration of a Bluetooth based ECG system into clothing. *Proceedings of the 8th International Symposium on Wearable Computers*, 186–87.

Park, S., and S. Jayaraman. 2001. Adaptive and responsive textile structures (ARTS). In T. Xiaoming (Ed.), *Smart fibres, fabrics and clothing*, 226–45. Woodhead Publishing Ltd. and CRC Press LLC.

Picard, R., and J. Healey. 1997. Affective wearables. *Personal Technologies* 1:231–40.

Post, E. R., and M. Orth. 1997. Smart fabric, or "wearable clothing". In *Digest of Papers of the 1st International Symposium on Wearable Computers*, 167–8. Los Alamitos, CA: IEEE Computer Society.

Post, E. R., M. Orth, P. R. Russo, and N. Gershenfeld. 2000. E-Broidery: Design and fabrication of textile-based computing. *IBM Systems Journal* 39 (3/4): 840–60.

Post, E. R., M. Reynolds, M. Gray, J. Paradiso, and N. Gershenfeld. 1997. Intrabody buses for data and power. In *Digest of Papers of the 1st International Symposium on Wearable Computers*, 52–55. Los Alamitos, CA: IEEE Computer Society.

Randell, C. 2001. Computerised clothing will benefit textile manufacturers. *Technical Textiles International* 10 (7): 3–27.

Randell, C., I. Anderson, H. Muller, A. Moore, P. Brock, and S. Baurley. 2005. The sensor sleeve: Sensing affective gestures. In *Workshop Proceedings on Body Sensing, 9th International Symposium on Wearable Computers (ISWC)*, 18–21. Osaka: Osaka Castle.

Randell, C., and H. Muller. 2000. The shopping jacket: Wearable computing for the consumer. *Personal Technologies* 4 (4): 241–44.

Raskovic, D., T. Martin, and E. Jovanov. 2003. Medical monitoring applications for wearable computing. *The Computer Journal* 47 (4): 494–504.

Robinette, K. M., and J. J. Whitestone. 1994. The need for improved anthropometric methods for the development of helmet systems. *Aviat. Space Environ. Med.* 65 (4): A95–99. Alexandria: Aerospace Medical Association.

Rocha, L. A., and J. H. Correia. 2006. Wearable sensor network for body kinematics monitoring. *Proceedings of the 10th International Symposium on Wearable Computers (ISWC)*, 137–38.

Sawhney, N., and C. Schmandt. 1998. Speaking and listening on the run: Design for wearable audio computing. In *Digest of Papers of the 2nd International Symposium on Wearable Computers*, 108–15. Los Alamitos, CA: IEEE Computer Society.

Smith, J. 2007. Communicating emotion through a haptic link: Design space and methodology. *International Journal of Human-Computer Studies* 65:376–87.

Solaz, J. S., J. M. Belda-Lois, A. C. García, R. Barberà, J. V. Durá, J. A. Gómez, C. Soler, and J. Prat. 2006. Intelligent textiles for medical and monitoring applications. In H. R. Mattila (Ed.), *Intelligent textiles and clothing*, 369–98.

Starner, T., D. Kirsch, and S. Assefa. 1997, The Locust Swarm: An environmentally powered, networkless location and messaging system. In *Digest of Papers of the 1st International Symposium on Wearable Computers*, 169–70. Los Alamitos, CA: IEEE Computer Society.

Stein, R., S. Ferrero, M. Hetfield, A. Quinn, and M. Krichever. 1998. Development of a commercially success wearable data collection system. In *Proceedings of the Second International Symposium on Wearable Computers*, 18–24. Los Alamitos, CA: IEEE Computer Society.

Sung, M., K. Baik, Y. Yang, J. Cho, K. Jeong, and G. Cho. 2007. Characteristics of low-cost textile-based motion sensor for monitoring joint flexion. In *Proceedings of the 11th International Symposium on Wearable Computers—Student Colloquium Proposals*, 29–31.

Sung, M., R. DeVaul, S. Jimenez, J. Gips, and A. S. Pentland. 2004. Shiver motion and core body temperature classification for wearable soldier health monitoring systems. In *Proceedings of the 8th International Symposium on Wearable Computers (ISWC 2004)*, 192–93.

Swallow, S. S., and A. P. Thompson. 2001. Sensory fabric for ubiquitous interfaces. *International Journal of Human-Computer Interaction* 13 (2): 147–59.

Taccini, N., G. Loriga, A. Dittmar, R. Paradiso, and S. A. Milior. 2004. Knitted bioclothes for health monitoring. In *Proceedings of the IEEE Engineering in Medicine and Biology Society (EMBC)*. San Francisco, USA.

Taelman, J., T. Adriaensen, A. Spaepen, G. R. Langereis, L. Gourmelon, and S. Van Huffel. 2006. Contactless EMG sensors for continuous monitoring of muscle activity to prevent musculoskeletal disorders. In *Proceedings of the 1st Annual Symposium of the IEEE Benelux Engineering in Medicine and Biology Society Symposium*, 223–26.

Tan, H. Z., and A. Pentland. 1997. Tactual displays for wearable computing. In *Digest of Papers of the 1st International Symposium on Wearable Computers*, 84–89. Los Alamitos, CA: IEEE Computer Society.

Tao, X. 2001. Smart technology for textiles and clothing. In T. Xiaoming (Ed.), *Smart fibres, fabrics and clothing*, 1–6. Woodhead Publishing Ltd. and CRC Press LLC.

Tao, X. 2005a. Introduction. In T. Xiaoming (Ed.), *Wearable electronics and photonics*, 1–12. Woodhead Publishing Ltd. and CRC Press LLC.

Tao, X. 2005b. Wearable photonics based on integrative polymeric photonic fibres. In T. Xiaoming (Ed.), *Wearable electronics and photonics*, 136–54. Woodhead Publishing Ltd. and CRC Press LLC.

Tian, X., and X. Tao. 2001. Mechanical properties of fibre Bragg gratings. In T. Xiaoming (Ed.), *Smart fibres, fabrics and clothing*, 124–49. Woodhead Publishing Ltd. and CRC Press LLC.

Toney, A., L. Dunne, B. H. Thomas, and S. Ashdown. 2003. A shoulder pad insert vibrotactile display. In *Digest of Papers of the 7th International Symposium on Wearable Computers*, 35–44. Los Alamitos, CA: IEEE Computer Society.

Tröster, G. 2004. The agenda of wearable healthcare. In R. Haux and C. Kulikowski (Eds.), *IMIA yearbook of medical informatics 2005: Ubiquitous health care systems*, 125–38. Stuttgart: Schattauer.

Van Langenhove, L., and C. Hertleer. 2004. Smart clothing: A new life. *International Journal of Clothing Science and Technology* 16 (1/2): 63–72.

Whitestone, J. J. 1993. Design and evaluation of helmet systems using 3D data. In *Proceedings of the Human Factors and Ergonomics Society 37th Annual Meeting*, 63.

Wijesiriwardana, R. 2006. Inductive fiber-meshed strain and displacement transducers for respiratory measuring systems and motion capturing systems. *IEEE Sensors Journal* 6 (3): 571–79.

Winchester, R. C. C., and G. K. Stylios. 2003. Designing knitted apparel by engineering the attributes of shape memory alloy. *International Journal of Clothing Science and Technology* 15 (5): 359–66.

Yang, D., X. Tao, and A. Zhang. 2001. Optical responses of FBG sensors under deformations. In T. Xiaoming (Ed.), *Smart fibres, fabrics and clothing*, 150–73. Woodhead Publishing Ltd. and CRC Press LLC.

Yang, Y., M. Sung, J. Cho, K. Jeong, and G. Cho. 2007. The influence of bending and abrasion on the electrical properties of textile-based transmission line made of Teflon-coated stainless steel yarns. In *Proceedings of the International Conference on Intelligent Textiles,* 11–13.

2 Designing Technology for Smart Clothing

Joohyeon Lee, Hyun-Seung Cho,
Young-Jin Lee, and Ha-Kyung Cho

CONTENTS

2.1 DESIGN PROCESS FOR SMART CLOTHING

2.1.1 THE NECESSITY OF A NEW DESIGN PROCESS FOR SMART CLOTHING

Smart clothing is a new type of apparel created by using a fusion technology that combines electronic engineering and apparel design. Since smart clothing requires a combination of several generically different features such as electronic efficiency, electrical safety, physical comfort, and aesthetics of a garment, the designers working on smart clothing should consider the multi-faceted factors in their design process. Even with such an interdisciplinary vision, however, it will be difficult for the apparel designers to avoid confronting fatal limitations in reflecting that vision in their design process, so far as they apply the traditional apparel design process to the creation of an apparel line of smart clothing, because none of the steps for such multivariate consideration has ever been included in the traditional process. Therefore, for successful designing of smart clothing, it is prerequisite to modify the traditional apparel design process to include some appropriate steps for interdisciplinary creation. In this chapter, we illustrate the latest models of newly modified apparel design process for smart clothing, and introduce the major results from a research project, Technology Development of Smart Wear for Future Daily Life, carried out from 2005 to 2009 with the support of the Ministry of Knowledge and Economy of the Korean government.

2.1.2 DESIGN NEEDS FOR SMART CLOTHING

In all the fields of industrial design, "designing" means a process which starts from an analysis of design needs and ends at their synthesis into some visualized forms, where "design needs" imply expectations held by consumers and manufacturers toward the object to be designed in the contexts of aesthetics, functionality, ergonomics, safety, and price. In short, design needs indicate what should be harmoniously reflected in every single "designed" product as conceptually fundamental composites. In other words, it means that critical changes occur in the design process when some new design needs emerge. In the case of smart clothing, the territory of design needs for apparel should be widened to include new categories, such as embedding of digital functions or the interaction between parts of digital function and human body.

In the field of smart clothing, some researchers have analyzed design needs (or requirements) according to their varying definitions. Defining smart clothing as "wearable motherboard" in her analysis on "GTWM" (e.g., "Wearable Motherboard" of Georgia Tech), Tao (2001) has categorized the requirements into functionality, connectability, durability, maintainability, usability in combat, manufacturability, wearability, and affordability. Viewing smart clothing as a garment system of wearable technology, Dunne, Ashdown, and McDonald (2002) have analyzed the needs into the matters of thermal management, moisture management, mobility, flexibility, sizing and fit, durability, and garment care. Cho (2004) has defined smart clothing as digital clothing and categorized the requirements into durability, easy care, comfort, safety, and aesthetic satisfaction.

This section considers the implications of smart clothing in the context of three definitions. First, it is a type of clothing that should satisfy the basic needs for a garment and the wearer's emotion and sensibility. Second, it is a new type of clothing in which digital or mechanical function is embedded. Third, it is a clothing-typed human-machine interface where all the needs for the clothing system, the machine system, and interactions between them should be reflected harmoniously. These three definitions of smart clothing imply that multiple aspects including the user's subjective expectation for apparel, demands for the mechanical function, and comfortable interaction between user and the system, should be considered as fundamental needs in the entire stages of the design process.

Based on previous examinations of design needs, we suggest a total of 10 key categories in design needs for smart clothing: functionality, usability, comfort, maintainability, easy garment care, manufacturability, safety, wearability, durability, and appearance. Comfort, wearability, easy garment care, and appearance are the matters relevant to the definitions of "a type of clothing," whereas functionality and maintainability are related to the definition of "clothing with digital/mechanical function." Durability and manufacturability are the design needs associated with both definitions. Usability and safety are related to clothing-typed human-machine interface.

2.1.3   ESSENCE OF THE DESIGN PROCESS FOR SMART CLOTHING

Viewed as a mental process, a design process involves a set of highly organized procedures of problem solving where various types of information are collected and synthesized into a consistent concept and finally transferred into a visual form. As every case of problem-solving process does, design process varies along with the feature and structure of design issues to be solved. Smart clothing is a good example of where such an innovation in the design process should occur, because it is based on a set of design needs that combine the design needs for clothing with the design needs for digital products.

Smart clothing as a product made by integrating digital devices with clothing is different from general clothes in its components, and therefore it requires adoption of the IT (information technology) product development process for its design. Under this assumption, Lee (2006) has developed a smart clothing design process after organizing a tentative design process model for smart clothing and testing the efficiency of this model, applying the general theories for design process to her own experience in case studies.

After considering the general theories for design process, the traditional clothing design process theory (Regan et al. 1998), the sports wear design process theory (Kim 1999), and the product design process (Ulrich 2004), Lee (2006) first reviewed and compared the apparel design process with that of other commercial products, matching the similar process steps. A smart clothing design process model was derived by substituting the result of actual case studies into appropriate place of the workflow according to time order. It was found that the process of smart clothing design contains a stage that corresponds to the "system level design" of the industrial product design process. This is the stage that sets up a design plan about an experimental prototype organized in the previous stage, finds the related subsystem and interface, and modifies the prototype design. In the traditional clothing design process, this stage is almost always omitted.

While the traditional design process enters the implementing stage as soon as a design concept is established, the smart clothing design process first composes an experimental prototype and finds product architecture, subsystem, and interface, and modifies the design. Only then, it enters the design implementing stage just as in the industrial product design process. In addition to this difference, the smart clothing design process

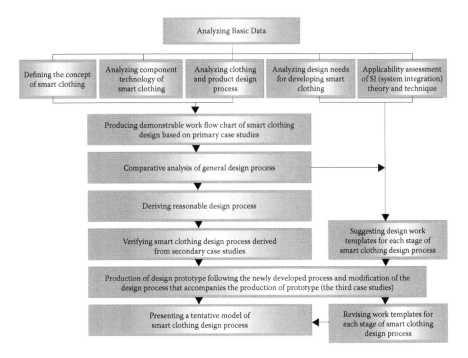

FIGURE 2.1 Flow chart of development of smart clothing design process model. (From Lee, Y. 2006. A model of design process for digital-color clothing. Ph.D. dissertation, Yonsei University.)

must include a testing step for reliability and performance, prior to estimating the step about the final prototype, unlike the traditional clothing design process.

2.1.4 Design Process Model for Smart Clothing

Lee (2006) has actually produced the smart photonic clothing by deriving a tentative model of smart clothing design process passing through a series of connected processes as shown in Figure 2.1, and applying the tentative model. The final plan of the smart clothing design process was drawn by analyzing the development process after complementing and revising it. The work template for various types of design to support every step in the design process was developed first. It was then offered to the designers so that they could execute the whole process of smart clothing design by simply following what the work template requested of them. The designers verbally reported details in their work process in order by carrying out "Concurrent Protocol Method" (a way of progressing design and at the same time interviewing about intention and content of a stage of each design) and "Retrospective Protocol Method" (a way of interviewing designers finishing the design about the intention and content of a stage of each design with contextual inquiry, showing the design works photographed) side by side. Then, a work flow chart of smart clothing design was produced. The designers who executed the design and participated in interviews

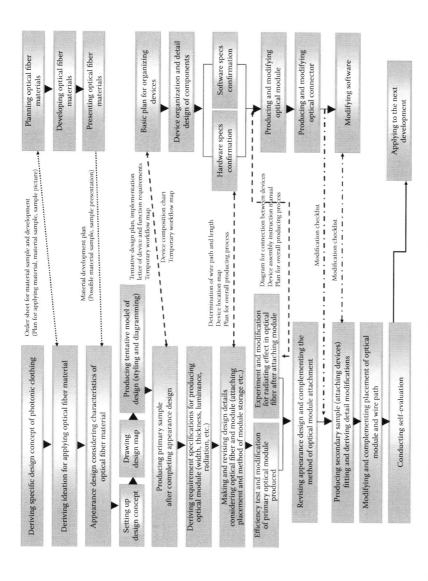

FIGURE 2.2 Design process model for smart photonic clothing.. (From Lee, Y. 2006. A model of design process for digital-color clothing. Ph.D. dissertation, Yonsei University.)

were five experienced people who have been engaged in developing smart clothing (Figure 2.2).

The design process illustrated in Figure 2.2 is the design process for optical fiber–based smart clothing known as photonic clothing. Lee (2006) stipulated that the system-level design step for this particular design process should be divided into two sub-steps: the one related to the controller part (PCB, wire path, connector, power unit) and the other to the output part (optical fiber material). The design prototype should also be developed separately in first and second round design models in accordance with the two sub-step system level design above. When this design process is followed, the design team will determine at the initial stage the detailed specs of the optical fiber material appropriate for the design concept and place the purchase order. Then, based on the material specs, the team will establish the draft appearance design of the clothing, and set the basic devices planning for the design and place the devices order. By combining all these, the team develops the first round smart clothing model (phonic clothing). At the next stage, by implementing garment fitting and midterm design review using the first round clothing model, the appropriate device placement and size, wire length and path, specific device and signal transmission-line planning, the team assesses functionality and makes device modifications. The team also produces related appearance design and makes interior structure planning modifications, and then the team combines all these together and produces the second round clothing model.

Producing a first round design model with the focus on material characteristics and a second round design model with the focus on the fitting and devices part is similar to the mock-up production in the product design process, but it differs from the sample production in the general clothing design process, which is used to review the material and the fitting together.

2.1.5 Characteristics of Smart Clothing Design Processes

The characteristics of the smart clothing design process can be classified into three categories.

First, the system-level design step is added to the smart clothing design process. This is a very important addition because in general product design processes, the product structure and substructure system and connection installation are defined first for concept development in the system-level planning step, and the planning modifications are made at the initial step. A similar step is found also in the smart clothing design process. Such a step is omitted in the traditional clothing design process or inconspicuously included in the concept development step.

Second, there has to be efficient communication with the areas developing materials or devices. The smart clothing design process differs from the traditional clothing design process in that requests are conveyed at each design step and adjustments of such requests are made. Moreover, this process can be repeated without restriction. Since this process is repeated, there is a need to keep the record of the participant's opinions and work progress in detail.

Third, the design process should consider the clothing-focused design process and the device-focused smart clothing design process in parallel. For a smooth design process implementation, therefore, various forms are needed as supplementary tools for designers to think while working, so that each process can be compared and evaluated in good balance.

Design work templates for smart clothing design were developed to meet such needs, and the design work templates for each step are the designer's worksheets in which various information needed for each design step is entered: all the details of step implementations, various devices and signal transmission-line specs, assignment of materials, amounts of materials used, etc. The details are summarized below.

1. System Level Design I (device, appearance, idea development stage): In this step, conception of appearance design; composition of optical module, battery and other devices; and the method of emitting light are defined. For smooth implementation of the process, it is most important to agree on the composition of devices, and a design work template to satisfy this condition should be proposed.

2. System Level Design II (first mock-up development stage): Unlike the general clothing design process, the appearance design is completed first as a step for detailing and determining the appearance design. The first round sample is completed only after the various devices and connection specs and length are derived.

3. Detail Design I (device detail design, development stage): After producing the sample of completed appearance design, this is the step in which the device and connection specs such as placement, size, and shape are determined based on the sample. Through this step, a design work template is created which minimizes trial and error and reduces later modification work by increasing the accuracy of device specs.

4. Detail Design II (interior design and appearance design modification stage): Assuming that the appearance design is modified and device specs are established, in this step the clothing design is modified and enhanced, and the interior designs for device storage pocket, wire paths, and input apparatus application method are designed in detail. To help understand the device and wire composition, 3D diagramming technique is introduced.

5. Midterm Design Review (design check stage): Before producing the second round sample, the sample is reviewed using a checklist to see if there are any design needs omitted in terms of clothing and device composition, and if there are erroneously designed parts. If items are given a "No" in the checklist, they are returned to the preceding step for modification work.

6. Modification and Complement (performance and fitting check stage): This is the step where, after the second round sample is completed, functionality and body fitting is performed with the devices and wires all attached to assess for device errors and design modification areas, and modifications are made as needed. The design work template is created to allow exchange of feedback with the developer of the materials and the devices, regarding problems related to the development.

7. Designer's Worksheet: Designer's Worksheet presents directions for the final work, stating necessary information for producing previously completed design, namely, specifications of every kind of device and signal transmission-line, material designation, and amounts of materials required and others.

In short, the design process model of smart photonic clothing has brought about a new concept of design process that tries to integrate the traditional design process for apparel with that of manufacturing industrial products.

This new concept in the design process model of smart photonic clothing emphasizes the system-level design, and the fitting and fine tuning procedures involve more extended stages for testing the functions of devices and their dynamic wearability. This new design process is expected to contribute not only to smart photonic clothing development but also to the design of other smart clothing.

2.2 PLANNING AND DESIGNING FOR SMART CLOTHING DEVELOPMENT

As a link for the research project "Technology Development of Smart Wear for Future Daily Life," the Smart Wear Research Lab of Yonsei University developed an MP3-playing jacket for entertainment; photonic clothing which is designed to radiate light for the jacket, applying an optic fiber and LED (light-emitting diode) module; and bio-monitoring cloth which can measure various vital signs for health care.

This section considers mainly the components of the newly developed smart clothing and the technologies involved in their planning and design.

2.2.1 MP3-PLAYING JACKET

Smart clothing with MP3 functions is a type of entertainment smart clothing that allows the wearer to use an MP3 player hands-free, manipulating the controller attached to the interior or exterior of the apparel.

We first considered the components and operation principles of an MP3-playing jacket and presented the development of a commercialized model of an MP3-playing jacket, through cooperation between the Smart Wear Research Lab, Yonsei University, and Beaucre Merchandising Inc., South Korea.

2.2.1.1 Components and Operating Principles of an MP3-Playing Jacket

The key components of an MP3-playing jacket are an MP3 player embedded or attached to the jacket, a controller to operate the MP3 player, a signal transmission-line connecting the controller and the MP3 player, an earphone, an earphone line, and the jacket on which these are mounted. If needed, a connection board between the MP3 player and the controller can also be added.

When this jacket is viewed as a system, the MP3 player functions as the CPU, the jacket's controller as the input unit, the earphone as the output unit, the earphone line and controller connection as the interconnection, and the jacket as the platform.

When the user inputs the signal (on/off, backward/forward, volume up/down, pause) to activate the MP3 player via the controller on the jacket, this signal goes through the signal transmission-line and reaches the MP3 player. Then the MP3 player is activated, and the sound signal generated from the MP3 file is transmitted to the earphone through the earphone line and the sound is produced. The connection board between the MP3 player and the controller, needed when using an MP3 player without loaded remote control support, is a control board added to support the remote control functionality of a type of remote controller. This controller attached to the jacket activates the MP3 device.

For most MP3-playing jackets, the signal transmission-line of controller connection has been developed on conductive fabric or yarn with relatively low conductivity because it is possible to output the signal with no significant impact on sound quality without high-grade signal transmission-line since the sound signal of MP3 file is generally in the low frequency of 200–250 MHz.

2.2.1.2 Types of MP3-Playing Jacket

The latest smart clothing developed worldwide is mainly categorized into two types: the type with an MP3 player embedded in the clothing and the type with a detachable MP3 player.

1. Clothing with an embedded MP3 player: Infinion Technologies in Germany, in cooperation with Munich's design school, developed an MP3-playing jacket embedded with a waterproof MP3 player device. Inside this jacket, a microcontroller supporting MP3 function, low-power consuming sound processing chip, multimedia card, battery, earplug, and flexible sensor keyboard are connected through conductive fiber material; the microcontroller and sound procession chip are contained in a waterproof container and embedded in the jacket. This jacket can be washed, and is designed to have almost no difference in appearance when compared with usual apparel. With the use of a miniaturized semiconductor chip with MP3 player functions, the sound source can be stored and exchanged by replacing just the memory card, without detaching or mounting the MP3 player (www.interactive-wear.com).

2. Clothing with a detachable-type MP3 player: The core technology of this type of MP3 smart clothing was developed most notably by Elecksen in England. Using its core technology called ElekTex, the company developed an electro-conducive smart fabric touchpad. As an input device based on light and flexible fabric, this product is durable and machine washable, and has five buttons to support MP3 functions. It is designed to be compatible with Apple's MP3 player, the iPod, through the interface box in which the connection board is known as the "iPod controller." For uploading, charging, or laundering, the earphone and iPod controller must be detached, and it is possible to launder the fabric touchpad and the fabric-based signal transmission-line together with the clothing. Such fabric-based touchpad technology is applied and commercialized in a variety of products through cooperation with many world-renowned apparel companies (www.eleksen.com).

FIGURE 2.3   MP3-playing jacket "W.α." (Courtesy of Beaucre Merchandizing, Inc.)

2.2.1.3 Development Case of Commercialized Model of MP3-Playing Jackets

In cooperation with the Smart Wear Research Lab, Yonsei University, Korea's Beaucre Merchandizing Inc. developed the MP3 function smart jacket W.α (Figure 2.3). In November 2007, this line was simultaneously introduced in two styles at 70 W. shops in Korea, China, Taiwan, and Vietnam.

W.α was conceived as a commercialized product line of Yonsei University's detachable type MP3 function clothing technology, with iPod compatibility attained through a license agreement with Apple Inc. This was designed to enable iPod control manipulation from the exterior part of the apparel by connecting the keypad embedded in the garment to the iPod with a connector. The keypad on the jacket's exterior can be manipulated without taking out the connected iPod from the pocket or bag, and since the devices (keypad, signal transmission-line) embedded in the jacket are all made of fabric, machine washing is possible after the connector box and the iPod are detached (Figure 2.4).

2.2.2 PHOTONIC CLOTHING

Photonic clothing is a general term for clothing that has an embedded or detachable illuminator and a digital control device emitting light while preserving the characteristics of the conventional clothing (Lee 2006).

Due to the chemical properties of clothing materials, the existing clothing undergoes varying degrees of fabric discoloration in accordance with temperature and light of the surrounding environment. By contrast, however, the photonic clothing enables wearers to control the degree of illumination with a light-emitting device.

2.2.2.1 Types of Photonic Clothing

Photonic clothing that has been developed so far can be classified into two types depending on the kind of illuminator. One is plastic optical fiber (POF)-based

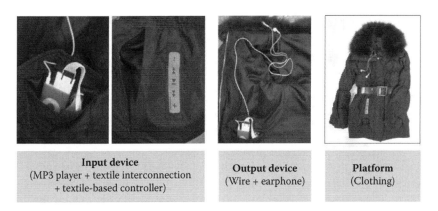

Input device (MP3 player + textile interconnection + textile-based controller)	**Output device** (Wire + earphone)	**Platform** (Clothing)

FIGURE 2.4 Components and clothing design of an MP3-playing jacket "W.α." (Courtesy of Beaucre Merchandizing, Inc.)

photonic clothing. It has LED-created digital color reflected on the surface of fabric through POF in it. The other type of photonic clothing uses electro luminescence (EL) as an illuminator and as a means of expressing light as well. There are other ways of applying photonic fabrics to clothing on the basis of characteristics of each illuminating material.

- POF-based photonic clothing: POF-based photonic clothing technology enables the transmission of light by manipulating a refractive index profile of the sectional structure of optical fiber. Optical fiber can promptly transmit a vast amount of data with ease. Lighter than copper cable, it has little leakage of information while transmitting data en masse. As thin as a strand of hair (approximately 0.1-mm diameter), it can be woven into a type of cable that binds various kinds of fiber (Cho 2004).
  - Optical fiber can be divided into two types: glass optical fiber made of pure quartz glass, and plastic optical fiber (POF). Glass optical fiber (GOF) is of superior performance but expensive. POF has the advantage of being relatively inexpensive and easy to handle, while its performance is lower than that of glass optical fiber. Due to its capability of transmitting a large amount of data with a minimal loss of light, optic fiber has been used mainly for telecommunications. Its function of transmitting light has been applied to the production of decorative illuminators, electronic scoreboards, condensers, and optical receivers.
  - An examination of the cross section of optical fiber shows its main structural components: the core that transmits light, and the cladding that has a function of cutting off light by surrounding the core, both forming double cylinders. The core is made of transparent material with a high refractive index, but the cladding is a cylinder with a relatively low refractive index. Optical fiber is created by applying a phenomenon known as total internal reflection where light is transmitted

from one point to another if light hits at certain angles on the surface of two materials with different refractive indexes.

- Throughout the world, optical fiber–based photonic clothing has been developed for the purpose of transmitting signals and producing digital color clothing. In the case of photonic clothing aimed at signal transmission that has been developed with GOF, it has been reported that the material's limitation in bending the radius hindered the production of fabrics blended with other threads. On the other hand, digital color clothing is mainly based on POF. Its principle of automatic color-realization is as follows: Colors of this type of photonic clothing are created by combining a light source with a medium of light transmission. In other words, the light automatically emitted from the illuminator by means of electric signals leaks through the etched surface of POF used as a light transmission medium.

- Photonic clothing applying EL: Another technology tapped for the production of photonic clothing is EL (electro luminescence), a future technology for the advanced generation display. The existing organic EL and other inorganic EL are all included in EL technology.

  - Depending on the inclusion of carbon, EL is classified into organic and inorganic types. Organic EL is a carbon-based chemical compound of hydrogen, nitrogen, and oxygen, while inorganic EL does not contain carbon. Organic EL is self-glowing. It emits light from the entire surface, so that brightness remains even regardless of viewing angles. Lighter and more flexible than the conventional LEDs with back light, EL is used in LCD back light of small-size terminal devices such as cell phones and PDAs. It also consumes less power than LED. Characteristically, however, exposure to the outside reduces its life cycle. Since it has a low tolerance to moisture, a protective measure—surface coating or blocking—should be taken. Because of the problems with its life cycle, organic EL is not used in devices that are operated for long hours. In contrast, inorganic EL is considered highly safe because even exposure to moisture or atmosphere is unlikely to cause its decomposition or transformation. It is resilient and has a long life cycle. It is known to have high heat resistance and emits little heat. It consumes one third less power than LED. With the entire surface glowing, it emits light more evenly than LED, which has a small-area light source. Newly developed inorganic EL, which has higher luminance than organic EL, glows clearly even in the daytime. It takes twice as much time as its organic counterpart to have brightness reduced by half at the end of its life span. EL is created by silk screening each layer of ITO film with pigments.

  - Inorganic EL, an illuminator activated by electricity, has been used in cell phone keypads and other interior products that consume a low level of power. Easy to break and bend, and sensitive to moisture, however, it has had only limited applications in household goods including clothing and bags. Thanks to the recent progress in technology development, a new technology of developing inorganic EL has thus allowed its application to photonic clothing in various ways.

2.2.2.2 Development of Smart Photonic Clothing

The photonic garment developed by The Smart Wear Research Lab, Yonsei University, has advantages of out-glowing similar garments by more than 30% luminance, and providing diverse functions and colors in accordance with wearers' condition and preference.

The core development technologies of photonic clothing are the manufacturing technology for optical fiber clothing; technology developments for optical fiber weaving, light-emitting diode, and connection; and module development. Specifically, it has been tried to find productivity improvement by developing optical fiber materials for clothing subsidiaries (Figure 2.5), and Yonsei's photonic garment realizes high efficiency (a standard of brilliance) by improving the management of optical fiber material and device power. Currently, a commercialized model of photonic garment is being promoted in collaboration with researchers in diverse technology fields.

1. Color-changing photonic clothing: This type of photonic clothing has a function that enables the wearer to choose and control colors and patterns of the garment's luminescent part by regulating the mode of light emission on the optical fibers (Figure 2.6). The application of this comparatively simple function is suitable for functional sports clothing and safety garments. As

FIGURE 2.5 Various subsidiary materials using optical fiber (optical fiber tape, blade, piping, etc.).

FIGURE 2.6   Color-changing photonic clothing.

shown in Figure 2.6, letters and patterns which the wearer desires to create can be diversely expressed by applying them in the form of piping or dobby and jacquard tape to a variety of optical fiber fabrics.

2. Color-responding photonic clothing: The garments shown in Figure 2.7 are the kind of photonic clothing with colors that automatically vary with surrounding colors or that allow wearers to adjust them as they please. The photonic clothing with color recognition capabilities consists of a bare garment made of optical fiber and other fibers, a color sensor that perceives colors surrounding the clothing, RGB LED that emits light of color recognized at the optical fiber end point, and a color control module. Colors that the wearer inputs with the color sensor change to colors similar to what the optical fiber of the photonic clothing has sensed. For example, when the wearer puts the color sensor to the yellow T-shirt his or her friend is wearing, the color is recognized and the color of optical fiber clothing turns to

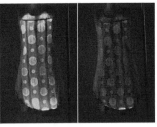

FIGURE 2.7   Color-responding photonic clothing.

FIGURE 2.8   Sound-responding photonic clothing.

yellow. The control device can even be programmed to recognize complementary colors and change colors of optical fiber accordingly.

3. Sound-responding photonic clothing: Figure 2.8 demonstrates photonic
clothing that reacts to sound. It is a kind of smart clothing capable of changing colors or light emission amount by reacting to external vibration or
sound. Smart clothing adjustable to external vibration consists of a bare
garment made of optical fiber and other fibers, a vibration sensor that perceives vibration external to the clothes, and a color light control module
that adjusts light emission of RGB LED to correspond to the level of vibration sensed on the optical fiber end point. In an environment such as a club
where music is present, the wearers are able to experience various effects,
with the area and the size of their garment's luminescent parts made to
change according to the loudness of external sound.

4. Photonic clothing applying inorganic EL for wearer's safety: Organizing
materials for developing a model of photonic clothing design on the basis
of inorganic EL technology consist of an inverter providing electric current
to emit light for inorganic EL, a stereo jack for connecting the inverter and
cable and signals delivery, a coated cable for connecting EL film and stereo
jack and signals delivery, a thermoplastic polyurethane film for improving durability of inorganic EL film, and a hot melt film functioning as a
thermoplastic adhesive. This garment was developed for walking at night
or for night activity such as an extreme outdoor sports wear. It also helps
the wearer's safety by ensuring the visibility range as EL embedded in the
clothing radiates (Figure 2.9).

FIGURE 2.9 EL smart photonic clothing.

2.2.3 Bio-Monitoring Clothing

2.2.3.1 Smart Clothing for Health Care

Viewed as ambient intelligence that is closest to the human body, the vital-signs-monitoring garment is a new kind of health-care clothing that blends biotechnology and information technology with fashion (Cho, Yang, and Sung 2008). In other words, this refers to clothing with embedded devices that can monitor vital signs, electrically converting the physical values measured from the body and expressing the resulting biological phenomena into electric signals. As such, it makes health management and remote treatment possible.

Vital-signs-monitoring clothing is constructed on a system that measures, analyzes, transmits, and feeds back the human body's vital signs. A variety of devices attached on the clothing can sense the vital signs of the wearers, such as breathing and heart rate, monitoring their health conditions round-the-clock. The so-called "Second Generation Smart Clothing Group," currently under development, is mostly "high-end" clothing with vital-signs-monitoring functions that require more sophisticated noise-control technology. The development of this clothing is based on a vision that looks forward to a lifestyle shift toward the concept of well-being and the trend of aged society that places much importance on health care. From this point of view, the need to innovate health-care methods has been addressed, and vital-signs-monitoring clothing, as one of such efforts, is regarded as a product of great potential in the development of smart clothing. At present, research and development of health monitoring clothes is actively under way throughout the world. The demand for health-monitoring smart clothing, including the type that has a system that senses and reacts to the environmental stimuli to the human body, is expected to increase as lifestyles become more diversified and interest in well-being increases.

2.2.3.2 Considerations in Designing Bio-Monitoring Clothing

Smart clothing is a new breed of clothing that combines machine and garment together. It should be present in the wearer's private space and controlled by the wearer. It should also maintain constancy in ready interaction with the wearer (Cho et al. 2000). Particularly for vital-signs-measuring smart clothing, the issue of constancy should be emphasized because it has to be constructed in consideration of the accuracy and stability of sensor signals, safe attachment of sensor into the garment, and the wearer's skin contact with a sensor. Sensor-based smart clothing should take different design approaches from other smart clothing because it has to capture vital signs of the human body with an embedded sensor.

From a medical engineering perspective, the sensor in bio-monitoring clothing should minimize energy extracted from the human body in measuring vital signs, make non-invasive contact with the living body system, and react only to energy reflected in the measured data. Therefore, a sensor unit of bio-monitoring clothing should be designed in such a way that the wearer will feel its presence as little as possible while it is integrated into the garment. The placement of devices should be selected in consideration of factors related to the human body (motion, the wearer's size and figure), factors related to devices (characteristics and limitations of sensor performance), and environmental factors (characteristics of context in the use of the garment).

The requirements of vital signs monitoring clothing include the following:

1. Biosensor-based smart clothing should have the vital signs reading unit of the sensor directly in contact with the skin, and yet it should not allow motions to affect the measurement of vital signs.
2. The problem with the usual smart clothing for vital signs monitoring is that body motions cause the electrodes to move, thereby generating noise. This can limit the measurement of the complex and highly functional signal system. Therefore, noise caused by motions should be kept to a minimum by designing smart clothing that can mitigate motion impact.
3. A person wearing clothing with a sensor embedded should not feel anything but the garment itself, in order to have a smooth monitoring of vital signs.
4. In addition, this clothing should be designed so that the wearers can interrupt vital signs monitoring whenever they want to.
5. This clothing should be designed in a way that it is as easy to wash and care for as conventional garments.

2.2.3.3 Development Case of Bio-Monitoring Clothing Prototypes

The Smart Wear Research Lab, Yonsei University, has developed a bio-monitoring smart clothing prototype that measures via ECG sensors exercise speed, ECG, pulse, and breathing.

This prototype's appearance design was made as a sports wear that has unobtrusive appearance, meeting the demand trend of customers, which is derived from previous research. Also it is designed not only to detach innerwear from outerwear but also to wear one innerwear with compatible outerwears, according to the wearer's need. A sensor for measuring vital signs, metal plated fabric (MPF)-based ECG sensor and device are embedded in the innerwear. A signal transmission-line for conveying signs is designed textile-based, and the device is detachable. The textile-

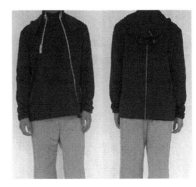

FIGURE 2.10 Design prototype of bio-monitoring clothing for measuring electrocardiogram. Left: Design for innerwear. Right: Design for outerwear.

based electrode is specially designed to minimize motion artifact so that the absolute placement of the electrode is not changed by movement (Figure 2.10).

This clothing prototype for health care, as a garment consisting of an ECG module and ECG electrodes, can continuously check ECG and thus has the function of aiding the prevention of heart attack and myocardial infarction. Measured ECG signs are transmitted wirelessly to the wearer's PDA, PC, and cell phone, so that the wearer can check his or her health condition. The scenario of basic model development consists of the following: In an emergency situation where the wearer is being taken to a hospital, a rescue worker can administer more professional and effective first aid after obtaining the patient's history and private data by reading the RFID tag attached to the wearer's arm through the RFID reader. The personal information will be transferred to the hospital so that emergency measures can be taken immediately on the patient's arrival.

Sensors and devices used in the model development include an ECG module, ECG electrodes, temperature sensor, velocity sensor, communication module for wireless communication, and power batteries. In addition, an RFID tag that stores the patient's history and personal data is embedded in the model.

The sensors and devices comprising this smart clothing for vital signs monitoring have four functional components: sensor, arithmetic, communication, and power source units. The sensor unit consists of a temperature sensor capable of measuring vital signs, and ECG electrodes. This sensor part is driven in connection with the other parts: the operating part aimed at accurate ECG measurement with filtering; the transaction part transmitting the operating part's output such as heart rate, breathing rate, and temperature; the output part sending this vital signs output to the cell phone or the main computer; and the power part ensuring smooth supply of power. Figure 2.11 shows a simulation for comparing ECG before and after exercise.

Furthermore, the design prototype of smart car-racing wear developed by the Smart Wear Research Lab, Yonsei University, has embedded a temperature sensor, a galvanic skin response (GSR) sensor, and a main sensor module (Figure 2.12). The GSR measurement is utilized as an index displaying the state of emotional awareness. It was designed to prepare for sudden dangerous situations by measuring the changes

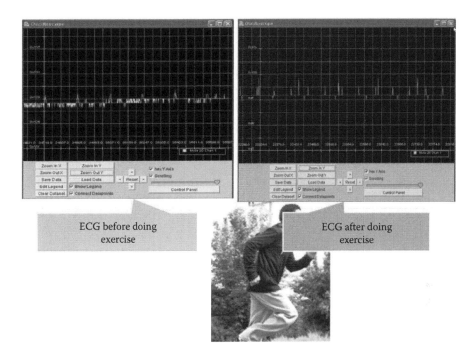

FIGURE 2.11 Simulation for comparing electrocardiogram before and after exercise.

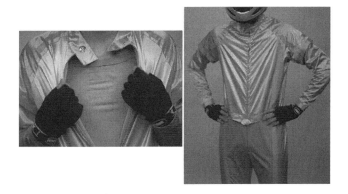

FIGURE 2.12 Design prototype for smart car-racing wear. Left: Innerwear. Right: Outerwear.

in temperature and sweating rate when the player's level of tension is getting high in racing. In other words, the rapid changes in the racer's temperature and the level of tension during training are measured through the embedded sensor and transmitted to an individual PDA or computer wirelessly throughout the main sensor module. Therefore, a coach can manage the amount of training, checking the player's health condition, before a health problem occurs. This clothing can assist more effective training because it is capable of grasping both psychological and physical conditions

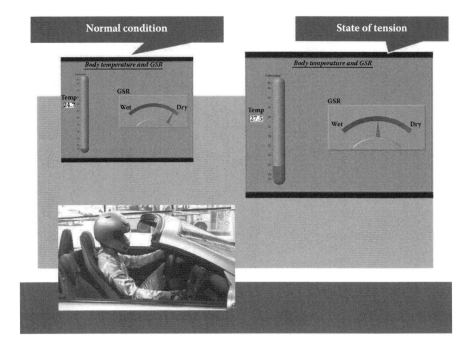

FIGURE 2.13 Simulation display for measuring the level of tension before and after racing.

of the player according to the situations in the race course while training. Figure 2.13 illustrates a simulation for measuring the level of tension before and after racing.

Since the devices in the smart car-racing wear are not exposed to the outside, there is no difference in the appearance from the conventional car-racing wear, and it is organized as a set of inner- and outerwears. The innerwear uses elastic materials, and because part of the sensor's electrode is designed to be in contact with the chest skin, and especially since the electroless plating fiber (EPF)-based sensor electrode is integrated, the level of tension and temperature can be measured. The temperature sensor module, the GSR sensor module, the main sensor module, and the battery are arranged dispersed in the outerwear, and they are all designed to be detachable. The transaction between the main sensor and the interface is designed to transmit signals through Zigbee-based wireless communication. The outerwear and innerwear are both washable when the modules and the battery attached to the outerwear are detached.

REFERENCES

Cho, G. 2004. *New High-Tech Textiles*. Seoul, Korea: Sigma Press.

Cho, G., J. Kim, W. Kim, M. Lee, and S. Lee. 2000. Digital clothing. *Fiber Technology and Industry* 4 (1): 148–57.

Cho, G., Y. Yang, and M. Sung. 2008. Development and its present status of bio-monitoring smart clothing and F-textiles. *Journal of the Korea Society of Clothing Industry* 10 (1): 1–10.

Dunne, L., S. Ashdown, and E. McDonald. 2002. Smart systems: Wearable integration of intelligent technology. In *Proceedings of the First Conference of the International Centre for Excellence in Wearable Electronics and Smart Fashion Products*. Cottbus, Germany.

Kim, Y. 1999. Sports wear design process. *Fiber Technology and Industry* 7 (4): 457–70.

Lee, Y. 2006. A model of design process for digital-color clothing. Ph.D. dissertation, Yonsei University.

Regan, C. L., D. H. Kincade, and G. Sheldon. 1998. Applicability of the engineering design process theory in the apparel design process. *Clothing and Textiles Research Journal* 16 (1): 36–46.

Tao, X. 2001. *Smart Fibers, Fabrics and Clothing*. CRC Press.

Ulrich, K. T., and S. D. Eppinger. 2004. *Product Design and Development*, 3rd. ed., McGraw-Hill.

3 Standardization for Smart Clothing Technology

Yong Gu Ji and Kwangil Lee

CONTENTS

3.1   INTRODUCTION

3.1.1   DEFINITION

In the next few years, clothing assortments are likely to incorporate more intelligence. The clothing will include the purposes of interaction, transporter, and interface for an enormously broad range of micro-systems. This latest innovation creates significant demands on creative capability in the clothing industry. Smart clothing is a combination of new fabric technology and digital technology, which means that the clothes are made with new signal-transfer fabric technology with installed digital devices. These clothing products will eventually be essential to the near future lifestyle. Since smart clothing is under development, there is no correct definition yet. So, it is often referred to as "digital smart clothing," "digital clothes," and "intelligent clothes." In this study, we define smart clothing as "new clothes that are convenient for use by IT-based applications" (Cho 2006).

3.1.2   MARKETABILITY AND PROSPECTS

The demand for high technical functions increases every day in the apparel market. The worldwide smart materials market is expected to rise at an average annual growth rate (AAGR) of 8.6% to $12.3 billion in 2010. The U.S. market for smart textiles was worth an estimated $70.9 million in 2006 and is expected to reach $391.7 million in 2012 (McWilliams 2007). However, due to the difficulty of quantifying the inputs of smart textiles required to produce such a wide range of products, many of which are still in a developmental stage, these market projections generally reflect the anticipated value of smart textile applications rather than the smart textile component per se. For example, projections for biomedical "smart shirts" are based on the market value of the shirt rather than the cost of the smart textile material used to fabricate it (McWilliams 2007). The clothing will merge into daily life and join the international fashion market by 2015 in the ubiquitous environment (Ahn 2004). Since 1910, there has been much change in the fashion world, but one constant value for clothing has been its function. Therefore, with the functions of clothing corresponding to the customer's demand, smart clothing is expected to be a major item for the future fashion industry. Smart clothing has been a high-value-added product with multi-functional materials since 1990, and this will rapidly grow in the fashion industry.

3.1.3 NECESSITY OF STANDARDIZATION

It is believed that competition among the United States, Europe, and other developed countries has begun in the international apparel market. In order to dominate the growing market, one must urgently focus on technology development. However, many problems have occurred due to the absence of standardization of technology. For instance, we currently manufacture all kinds of subsidiary materials to aid us in developing the technology. So it ends up creating many more obstacles for actually developing the technology. Therefore, the efficiency of technology development can be strengthened through industrial standardization. There are three benefits that can be gained from the standardization of smart clothing. First, smart clothing can reduce unnecessary expenses by increasing efficiency in product development. Second, this study can achieve a comparative advantage in technological competition. And third, customers will begin to trust their smart clothing, which can raise market shares in the fashion industry.

3.2 STANDARDIZATION AND PATENT TRENDS

3.2.1 GENERAL STANDARDIZATION TRENDS

Standardization can be accomplished in two ways: (1) taking leadership in the market, and (2) acquiring international standards such as ISO (International Organization for Standardization), IEC (International Electrotechnical Commission), etc. (Korea Industrial Technology Foundation 2007).

The general trend of the international standardization process is as follows:

- Search for the interested area.
- Promote the standardization process for about 3 years, including a technology development period.
- Exchange related information with people in charge of the ISO or IEC.

- Select it as the foreground standard, when an object is considered to have a high standardization effect.
- Compatibility standards, quality standards, and safety standards are growing issues of international standards.
- Seek ways to avoid each country's unnecessary technology regulations through the WTO (World Trade Organization) and FTA (Free Trade Agreement).
- Technological standards and normative laws will come together as laws are complemented with references to the technology standard.
- In the United States and Europe, standardization is normally achieved by private companies.
- In the United States, if a private institution for standardization approves the standard, then it is admitted into national standards.
- In the case of Japan, the government itself is the standardization institution.

3.2.1.1 Standardization Trends in the United States, Europe, and Japan

According to preliminary research, there was no application for ISO or IEC certification. Technology development and commercialization comparatively have been have been conducted with activity in the United States, Japan, and Europe. However, there are no studies or data related to standardization except in Japan, whose firefighting uniforms are certified by the IEC.

3.2.1.2 Standardization Trends in Korea

According to preliminary research, like other countries, standardization has not yet been implemented; there is only a standard for general clothing in Korea (Korean Standards 2008). However, the MKE (Ministry Knowledge Economy) is organizing a strategic technology project, which has commissioned the "e-textile development for daily life" research project aiming for function and process standards for the industry (Cho 2007). Early research results are expected to be available in the latter half of 2009 according to the interim report. Following this investigation of domestic and international trends, we were able to conclude that national standards for the products or technologies related to smart clothing have not yet been established.

3.2.2 Patent Trends

Patent trends need to be analyzed for smart clothing standardization. By researching patent classification as well as rise and fall rate of patents in each country, classification standard and priority of standardization category can be thoroughly analyzed. Patent analysis on smart clothing is accomplished based on "Technology and Market/Patent Analysis," sponsored by Korea Institute of Patent Information (Korea Patent Information Institute 2006). Through this information, it was possible to get better understanding of patent classification standards and the number of patent cases in each country. In the next phase, it will be used as base data for standardization studies.

3.2.2.1 Patent Definitions for Smart Clothing

Patents for smart clothing can be classified into four sections: (1) conductive textile materials, (2) phase change/discolor textile materials, (3) digital-devised clothing, and (4) medical clothing. The definition for each term is as follows (Korea Patent Information Institute 2006).

3.2.2.1.1 Conductive Textile Materials

Conductive textile materials are textiles with conductivity. These textile materials are used for intercepting electron waves or for a fuel cell's electrolyte film. Conductive textile materials let electrons and ions go through, just like an electric wire. Conductive textile material is basic and essential to the growth of smart-wear technology.

3.2.2.1.2 Phase Change/Discolor Textile Materials

Phase change textile material reacts to temperature change. It does not respond to a change in body temperature, but when there is a temperature change outside, phase change materials melt or solidify. Discolor material is also known as Chameleon material. It changes color when exposed to heat.

3.2.2.1.3 Digital Device Equipped Clothing

Digital-devised clothing involves an application of digital devices, such as music players or cell phones, to clothing. It is now regularly sold in many countries.

3.2.2.1.4 Medical Clothing

Medical clothing functions as telemedicine that provides real-time feedback to wearers, and alerts bystanders, if necessary, by applying a sensor or microchip to the clothing or textile product. It can be used to monitor the user's individual heartbeat rate, respiration rate, and temperature, or to record a soldier's wound status in a war, etc.

3.2.2.2 Patent Trends in Smart Clothing

In order to find out about patent trends in smart clothing, this study investigated the patent application and record status of Korea, Japan, Europe, and the United States. For Korea, Japan, and Europe, the investigation period was 1997 to 2003; for the United States, the investigation period was set from 1999 to 2005, because they do not open patent filing to the public for two years. We assumed the investigation time of patents pending to be 2–3 years. In the global patent occupancy share, Japan has 48% share, which is 762 cases out of 1606 cases; the United States has 23% share, which is 374 cases out of 1606 cases; Korea has 22% share, which is 358 cases out of 1606 cases; and Europe has 7% share, which is 112 cases out of 1606 cases. However, the United States only validates registered patents, which means that the number of total patent applications could be quite large (Korea Patent Information Institute 2006).

In each country, yearly patent applications in the smart textile field have increased since 2000. There has been no great change in the process of technology innovation in the smart textile field in Korea and Europe; however, according to increased interest in smart clothing, the number of patent application in Korea increased rapidly

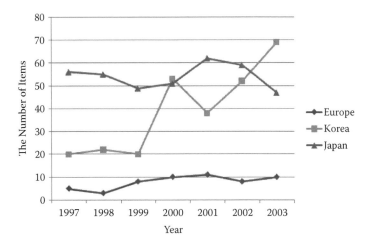

FIGURE 3.1 The number of patent applications in Korea, Japan, and Europe. (Modified from Korea Patent Information Institute. 2006. Smart textile technology and market. Patent analysis report. Seoul, Korea.)

in the year 2000. Japan showed some growth in the year 1999; however, in the year 2001, the growth declined. They got back on track in the decade before the year 2000. This is in contrast to Korea, where the number of patent applications constantly rose starting from 2001. Japanese patent applications are somewhat declining in number (Korea Patent Information Institute 2006) (Figure 3.1).

The United States does not open records of their patent applications to the public. So we separately considered 2–3 years for investigation time to patent filing. The United States showed a rapid decrease from the year 1990. There was some growth in the year 2001, but the growth declined again in the next year (Korea Patent Information Institute 2006) (Figure 3.2).

The smart clothing–related field, which includes medical clothing and digital-devised clothing, shares 48%, which is 772 cases out of 1606 cases, and the smart textile material field, which includes conductive textile materials and phase change/discolor textile materials, shares the remaining 52%, which is 834 cases out of 1606 cases. When categorizing smart textile materials into four parts, conductive textile material occupies 40%, which is 649 cases out of 1606 cases. Next, digital-device clothing shares 28%, which is 453 cases out of 1606 cases; medical clothing shares 20%, which is 319 cases out of 1606 cases; and phase change/discolor textile material shares 12%, which is 185 cases out of 1606 cases (Korea Patent Information Institute 2006) (Figure 3.3).

3.3   TRENDS IN TECHNOLOGICAL DEVELOPMENT AND INDUSTRY OF SMART CLOTHING

3.3.1   TRENDS IN TECHNOLOGY DEVELOPMENT

Technological innovations have been developed for use in easing daily life, fitness and medical support, and military and other special functional uses. Current developments in the United States are utilized for military, medical, and other specified work,

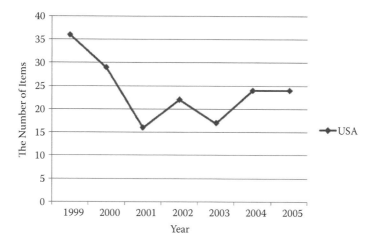

FIGURE 3.2 The number of patent applications in the United States. (Modified from Korea Patent Information Institute. 2006. Smart textile technology and market. Patent analysis report. Seoul, Korea.)

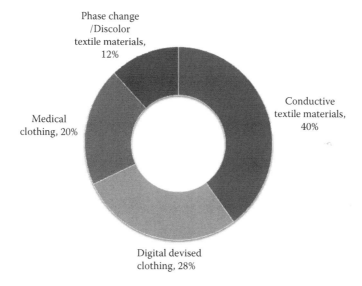

FIGURE 3.3 Patent application share per technology. (Modified from Korea Patent Information Institute. 2006. Smart textile technology and market. Patent analysis report. Seoul, Korea.)

rather than general day-to-day use. In contrast, Europe's main direction for developments is focused on fashion and consumer health. The devices attached to clothing are different depending on usage, but the role and requirements for e-textiles on clothing is generally the same. Current developments in South Korea smart-apparel development has involved a range of research institutes, universities, and conglomerates with the active support of the South Korean government. The government even has estimated that the global market for digital clothing could be $7 billion by 2014. South Korea wants more than 20% of it. (Korean Textile Development Institute Textile Information Team 2008).

3.3.1.1 Technology Development Features in Smart Clothing

- *An attempt to achieve ultimate mobility*: As people want to be empowered by electronic devices in order to access required information everywhere and every time (Marzano 2000).
- *Product miniaturization*: To achieve the highest level in terms of mobility, electronic devices are rapidly reduced in size. As a result, the personal electronic devices can be attached to clothes or become accessories. But by contrast, their functions and features continually increase. The conflict between function and size leads to difficulty in term of use, as the functionality embedded in electronic devices is often complicated and inaccessible (Van Heerden 2000).
- *Integrating IT devices into garments*: Integrating IT devices into garments might be an important element, since it provides many benefits such as mobility, which probably makes it easier to use (Busayawan 2003).
- *Advanced technological development*: These new technologies bring a large number of possibilities and opportunities for new applications. Many applications are the result of experiments with conductive properties of fabrics (Orth 1998).

3.3.2 Trends in the Smart Clothing Industry

According to the textile information team at the Korea Textile Development Institute (Korea Textile Development Institute Textile Information Team 2008), the current status of the industry is as follows.

3.3.2.1 Wearable Technology

According to the Eleksen company in England, smart textile commercialization has begun as a result of the increased demand for electrical products. Eleksen textiles have similar feeling and features to nylon, so they could be used on jumpers for wind protection, suitcases, or even a key chain. Moreover, this material does not require an electric wire or any other metal to run electricity through. Recently, Eleksen signed a contract with outdoor apparel company O'Neill to supply items for their H2 line.

The H2 line has a coat equipped with solar panels. This coat has a headphone and microphone, so there is no reason to take off the coat to listen to music or make a call when the mobile phone is inside the coat. A Bluetooth module installed in the product has the function to switch between a music player and phone. Eleksen's textiles are also used in a jacket and a back-supporting chair for the Spyder Active Sports Company.

The price of the coat is currently around $3,000, but the company is attempting to lower the price. To produce a medium range of coats, Eleksen is currently negotiating with a Taiwanese manufacturer of fabric keyboards and other PC peripheral devices to import back to the United Kingdom. The extra expense of textiles that conduct electricity inside clothing is estimated to be around $20. Altogether, it is estimated that the total market worth for these textiles will reach $500 million by 2008 (Harmon 2007).

The scientists who laid the foundations for this technology are Dan Sandbach and Chris Champman. The scientists began by developing a fabric to build puppets for a U.K. TV show called "Splitting Image." They required a material to allow the dolls to move smoothly. Following the show, they continued in fabric research, and in 1988 founded their own company (BBC Research 2007).

The technology they developed is a touch pad with five layers of raw materials, so if you touch it, all five layers will be pressed at once. Two surface layers are fabric and the middle one consists of "knit," so electrical current flows through it. Therefore, an electric current forms a circuit and sends a signal to a processor. This processor then transfers the contact information into the machine. So, for example, if a finger pushes "Play," the music starts. If you press the volume button, the music volume will change. Most of the clothes have resistance and each part is divided into sections. There is no need for a guidebook because a wearer will be able to recognize each clothing section's reaction to touch.

Eleksen's sales have recently been increasing. Turnover in 2004 was 0%; however, that turned into millions of dollars by 2005 (Reid 2006). The company will move on to the next stage with the technology already in place for commercial success. The first products from O'Neill equipped with this technology were winter sports clothes. The technology is also waterproof, so it can be used in a diving suit. In addition, Eleksen is planning to get into the business of producing a medical outfit, if there is a sufficient consumer market for the keyboard and coat (Comprehensive Merchandising Support 2007).

We conclude that Eleksen's products will make progress in technical development and commercial production. And also, we would expect success in the market because of the realistic cost of the products.

3.3.2.2   Japan Kuchofuku's New Air-Conditioned Shirt

The Japanese Kuchofuku Company says that it has a sensational (new, revolutionary) air conditioner to install in clothes. The air system in clothes retains cool air when it is hot. Two 10-cm-diameter small fans are installed in a shirt. The two fans are located on back of right waist and left waist. The fans are very light, there is no sagging, and they work with four rechargeable nickel or hydrogen batteries. According to Kuchofuku, the fans evaporate sweat by circulating air so the wearer can feel cool (Ishimaru et al. 2006).

This system is a development of the Japanese concept "Coolbiz." It all started when the Japanese government recommended workers wear cooler clothes to help cut down the use of air conditioners. However, according to the previous report, many workers found Coolbiz clothes uncomfortable to wear because they were not designed for daily life. Ichigaya Hirochi, a Japanese developer, has been working on an air-conditioned clothing design. He showed his first products in April 2005, including an air-conditioned long-sleeved jacket costing $104, which can be purchased online and through a catalog (Caferi 2007).

We conclude that Kuchofuku's products are very great in the aspect of technology. However, the clothes are uncomfortable to wear and the unit cost is high due to development costs. Therefore, we have to study technical development for users in special circumstances.

3.3.2.3   A Living Body Signal and Environment Monitoring Clothing

In Wearable Computer Laboratories, researchers have developed a three-part system to monitor heartbeat and the respiratory organs (Wearable Computer Lab 2008):

- Clothes combined with electrical engineering
- Connected wire combined with electrical engineering
- Computer built into clothes

The clothing collects data about the user's actions. Also, it is possible to wash the clothes without affecting any data or electrical parts. Metal bands of connected wires are built into the clothes, so the computer installed inside the clothes is connected with wires.

A touch screen is located on the surface of the clothes and each electrical material is able to transfer signals. The clothing will contain a cable for downloading any data stored on a computer. At the same time, this lead will be able to recharge the clothing for use the following day.

According to the researchers, such a technology could have a variety of uses in the monitoring services area. For example, a "monitoring service" makes it possible to check on Alzheimer patients staying at home. The new system also makes it possible to check the wearer's health. The wearer does not need to learn how to use the new machine but just needs to attach it on the clothes.

Bruce Thomas in West Austria, a professor working at Wearable Computer Laboratories, says the technology "monitors population statistics, weather forecast, and all activities for the day. But the most important thing for these items is their ability to check any disruption of activities such as falling, immobility or stopping eating. The technology can be categorized into two sections: general and non-general. It helps people who are living in deserted areas or in need of great attention from families. In case of emergency, medical treatment can be quickly assessed" (Wearable Computer Lab 2008).

Like this product, it has to merge with products that have a high degree of difficulty in technique and a variety of spearhead techniques. Therefore, it is essential to develop techniques continually and interchange with various fields. Also, as the products have relation to U-healthcare, demand will increase in the long term.

3.3.2.4   Music Player Clothing

Nowadays, music player smart clothes are under way, which is technical development, and now they are sold in not only overseas markets but also domestic markets. The important factors in future technical developments are compatibility to attach various music files, and comfort, and functionality to maintain music player ability. Next are the newest music player clothes.

3.3.2.4.1 *Zegna Sports' iJacket*

Ermenegildo Zegna is a worldwide brand. Zegna Sports has produced the iJacket, which has become a prominent Italian product for men. It is compatible with the Apple iPod. During routine chores or exercise such as hiking, the user is able to access the music player through a fabric keypad. This product also uses ElekTex's interface with Bluetooth so there is no need to reach for one's mobile phone. The fabric keypad on the outside of the jacket can be used to answer or hang up a call (Park 2007).

Because this product has characters that are comfortable to wear, and functions of music player and cell phone, we expect positive reactions from consumers.

3.3.2.4.2 *O'Neill's H3 Series, Comm. Emt Jacket*

Global sportswear company O'Neill is a major brand in the manufacturing of products for wakeboarding, surfing, snowboarding, etc. They have developed the Comm. Emt jacket for snowboarding, using ElekTex's technology (O'Neill 2008).

Techniques used by O'Neill are based on ElekTex's techniques, but the Comm. Emt jacket has big, special feathers, which are warm, waterproof, and also maintain the music player and cell phone in many environments like wakeboarding, surfing, snowboarding, etc. These factors show that music player clothes have to develop further to maintain ability not only in everyday life but also under various circumstances.

3.3.2.4.3 *Bagir's MusicStyle*

Global tailored suit brand Bagir is a revolutionary company that produces machine washable suits, waterproof suits with nano-technology, super-light suits, and more. They were the first to introduce the concept of music player functionality with ElekTex technology (Comprehensive Merchandising Support 2007).

It is the first time smart clothing technology has been applied to a suit. The development of music player clothes will expand to suits or fashion clothing from preexisting sports or casual clothes.

3.3.2.4.4 *Urban Tool's Groove Rider*

Austrian Urban Tool Company has already commercialized iShirts, which have a pocket made with elastic fabric for a mobile phone or iPod, so the devices won't slip out. A recently developed product is a shirt with a music player controlled by a smart fabric touchpad. The smart fabric touchpad is one of ElekTex's technology products (Comprehensive Merchandising Support 2007).

3.3.2.4.5 *Levi's Redwire DLX iPod Jeans*

The international American brand Levi's has added a music player on their symbolic blue jeans in contrast to tops where you are more likely to find music player items (Comprehensive Merchandising Support 2007).

The Levi's product, compatible with the iPod and other music player controllers, is attached on the right pelvis. The products are designed to keep an iPod in the thigh or back pocket. The headphone jack is designed to roll in for safekeeping (Comprehensive Merchandising Support 2007).

3.4 STANDARDIZATION METHODS

3.4.1 RESEARCH SCOPE

Smart clothing standardization scope can be divided into two main areas. In the first area is the clothing itself or the installed hardware sensor, digital device, chip, and accessories, and in the second area is the software protocol or system program to put into functioning the smart clothing. In this research, the focus was on the aspect of the standardization of hardware of smart clothing. Therefore, from the point of view of hardware, smart wear characteristics can be divided into wearable characteristics and electronic/electrical characteristics. Consequently, the research on the smart clothing standardization will be done taking into consideration these characteristics.

3.4.2 CLASSIFICATION OF PRODUCTS AND TECHNOLOGIES IN SMART CLOTHING

Classification of smart clothing and related technology is necessary for standardization. However, there is no absolute definition and no clarification about the related technology. To deal with these problems, this research analyzes the products and the technology of each institution in Korea. Table 3.1 is an analysis report on each institute's data (Korea Industrial Technology Foundation 2007).

As seen in Table 3.1, technology clarification has not yet been clearly established. Each institute has confused the technologies and used a mixture of information for development. This study implements smart clothing classification. Table 3.2 is the classification for smart clothing.

As suggested in Table 3.2, smart clothing can be categorized into four areas: body- and environment-monitoring clothing, IT device–equipped clothing, digital color clothing, and extra functional clothing. We defined the application, related technology, and common technology among these four categories. Smart clothing classification will be the first step towards a future standard. It will help to plan standardization schedules with technology priorities, marketability, urgency, and mutual uses.

3.4.3 STANDARDIZATION ANALYSIS OF SMART CLOTHING

We established the smart clothing standardization analysis according to the smart clothing classification. Clothing products that are selected are either in the research process or have a high potential to be commercialized. The selected clothing products are extracted as standardization items after collecting/analyzing related technology. Through the smart clothing standardization analysis, a total of 17 standardization factors, ranging from electrocardiogram (ECG) measurement factors to smart clothing terminology, were analyzed (Table 3.3). Extracted standardization items were ranked according to technology, marketability, urgency, and mutual operation standards.

To select standardization priority, each scale was classified into three steps and was applied to each of the standardization factors.

TABLE 3.1
Each Institute's Analysis of Smart Clothing

Institute	Type of Analysis Report	Classification
KIP (Korea Institute of Patent)	Smart textile analysis	Conductive textiles material area
		Phase change textiles material area
		Discoloration textiles material area
		Digitally devised clothes
		Medical clothes
KITF (Korea Industrial Technology Foundation)	Textiles industry technology road map	High functional high-tech textiles
		Intelligence smart textiles
		LOHAS (Lifestyles of Health and Sustainability) fashion clothes
		Future revolutionary textiles
KSTI (Korea Sewing Technology Institute)	Future fashion; "Smart clothing"	Leisure textiles
		Medical textiles
		Healthcare textiles
		Environment textiles
		Digital textiles
KMKE (Korea Ministry of Knowledge Economy)	Smart clothing commercializing technology business report	A living body signal and environment monitoring clothes
		Entertainment smart clothing
		Photonic clothes
		Extra functionality clothes

3.4.4 PRINCIPLE FACTORS FOR STANDARDIZATION

The six standardization factors (shaded parts of Table 3.3) that went through this process have the highest standardizing value. The selected standardization factors are as follows.

3.4.4.1 ECG Measurement Standardization

ECG measures and records the electrical activity of the heart in exquisite detail. Interpretation of these details allows diagnosis of a wide range of heart conditions; a standard for reliability, correctness, and detection rate should be established to develop smart clothing that supports ECG measurement.

3.4.4.2 Fabric Signal Wire Standardization

Fabric signal wire, which contains both an electrical quality and a fabric quality, is one of the main materials for smart clothing. The functional standard for washability, a clothing function, as well as electrical function, such as insulation resistance, should be established.

TABLE 3.2
Smart Clothing Classification System

Smart Clothes Category	Application	Related Technology	Common Technology
Signal detecting clothing	Body monitoring	ECG measurement PPG (photoplethysmography) GPS module GSR module	Signal sensor
	Bio-monitoring	Temperature/humidity sensor module Ultraviolet rays/ozone sensor module	
IT device–equipped clothing	Entertainment	Fabric signal wire Music player built-in Cell phone built-in Media built-in Fabric keypad Signal transmissibility materials	Conductive fabric
Digital color clothing	Fashion Life safety Entertainment POF fabric manufacture	Thin film electroluminescent technology Photoelectron weaving textile LED (light-emitting diode) RGB programming Color/sound responding sensor	
Extra functional clothing	Heat emitting function Air-conditioning in car racer uniform Masking in army uniform Thermal resistance for fire fighters Industrial safety clothing	Heat emitting system Air-conditioning function system Anti-radar Thermal resistance fabric RSI (repetitive strain injury) sensor	Digital convergence

3.4.4.3 Fabric Keypad Standardization

The keypad contains a series of buttons that are used for electrical functions. Fabric-based keypads can be defined as fabric containing keypad characteristics. Fabric keypads are able to transfer input signals through fabric signal wire. At the same time, they should be washable like other fabrics. Therefore, standardization for reliability and button input errors is necessary.

3.4.4.4 Photonic Textiles Module Standardization

Photonic textiles contain textile characteristics while using electrical energy to radiate various lights. Electric power efficiency and a safety standard should be established.

TABLE 3.3

Standardization Items and Priority Are Selected Based on Smart Clothes Products and Technology

Standardization Item	Standardization Type	Technology	Marketability	Urgency	Interchangeability
ECG measurement	Performance	●	●	●	●
Temperature/humidity sensor module	Development	◎	◎	○	○
GSR module	Development	◎	◎	○	○
Ultraviolet rays/ozone measurement module	Development	◎	○	○	○
Environment sensor	Performance	◎	○	○	○
Fabric signal wire	Performance	◎	●	●	●
Fabric keypad	Performance	◎	●	●	●
Signal transmissibility materials	Performance	○	●	○	●
Photonic clothes module	Product	●	●	◎	●
POF fabric	Performance	◎	◎	●	○
LED (light-emitting diode)	Performance	◎	◎	○	○
Photoelectron textiles materials	Development	●	○	○	○
Electron activity textiles materials	Development	●	○	○	○
Smart clothes interface	Performance	○	◎	○	○
Interface and components for smart clothes	Development	○	◎	○	○
Smart clothes usage test	Performance	○	○	○	●
Smart clothes terminology	Performance	○	◎	●	●

Note: Standardization priority: high (●), medium (◎), low (○).

3.4.4.5 Plastic Optical Fiber Fabric Standardization

Plastic optical fiber (POF) is the main material used in photonic textiles. As it could become the main technology for the smart clothing industry in the future, standards such as function evaluation and reliability of related technology should be established.

3.4.4.6 Related Terms for Smart Clothing

Currently there is no definite system of terms relating to smart clothing. Standardized terms are efficient and reliable when defining related materials or technology. Therefore, related terms should be standardized.

3.5 TEST EVALUATION FOR STANDARDIZATION

3.5.1 PERFORMANCE TEST EVALUATION AND RESULTS

One of the most important functions for smart clothing is that it contains textile characteristics and electrical/electrode functions. In accordance with this feature, the performance test evaluation method is divided into two parts. The first part is clothing property, which includes clothing compatibility, durability, materials, and related specifications. The second part is the electrical/electron property with the function and electrical characteristics related to digital device functions. The main reason for these tests is to evaluate methods of measuring electrical/electron properties as well as clothing functionality. Then, the result will be used for choosing standards that are necessary or applicable in the next phase. Several evaluation methods for standardizing general clothing have been referred by the KOFOTI (Korea Consumer Agency 2003). As for electrical/electric products, they refer to IEO/ISO/KS standard and approval tests. Researched function evaluations were tested by the KATRI (Korea Apparel Testing and Research Institute 2008) based on researched function evaluation methods.

3.5.1.1 Clothing Property and Evaluation

Smart clothing is basically considered as regular clothing. Therefore, it should have basic clothing functions such as size changes, appearance changes, and a heavy metal detection standard. An established standard for clothing products and standardization can be included in these standards for clothes.

We tested a sort and category evaluation for standardization. According to the Korea Apparel Testing and Research Institute (KATRI), the categories are appearance change, down and feather, mechanical properties, etc. Details are illustrated in Table 3.4. Among test evaluations for classes and categories, any factors suitable for evaluating will be used for determining characteristics and evaluation methods after verification (Korea Apparel Testing and Research Institute 2008).

3.5.1.2 Electrical/Electron Property and Evaluation

Smart clothing not only has the characteristics of clothing, but also the characteristics of electrical/electron elements. Therefore, we investigated electrical/electro-related test evaluation methods. Through research investigations, we could select test evaluations that are highly related.

For example, fabric electro buttons were tested and the evaluation was compared to previous research in findings by the ISO and IEC. A variety of connected wires, electrical resistance, insulation resistance, and R-peak detection were analyzed at test specifications. R-peak is acquired through measuring electrocardiogram signal (Table 3.5).

TABLE 3.4
Normal Clothing Evaluation

Category	Test Type	Relevancy	Importance	Possibility
Appearance change	Size changes rate	●	●	●
	Dry-cleaning size change rate	●	◎	●
	Steam press size change rate	◎	●	◎
	Iron size change rate	●	●	●
	Form change	●	●	●
Down and feather	Mixture rate	◎	◎	◎
	Oxygen	◎	○	○
	Smell	○	○	○
Mechanical properties	Breaking strength and elongation	●	◎	●
	Tearing resistance	●	●	●
	Abrasion resistance	◎	◎	◎
	Pilling resistance	○	◎	◎
	Seam strength	◎	●	●
	Flexibility	●	●	●
	Slip resistance	◎	○	○
	Wide fabric tensile strength	●	●	●
Fit and comfortability	Water repellency	◎	●	◎
	Water resistance	●	●	○
	Water vapor transmission rate	○	○	○
	Insulation	●	●	●
	Bending resistance	●	●	●
	Withstand voltage	◎	◎	●
	Durable press	●	○	○
Functionality	Germicidal (action)	○	○	○
	Mold resistance	○	◎	◎
	Far-infrared release	◎	●	●
	Infrared reflectance	◎	○	○
	Anionic release	○	○	○
	Fire-resistance	●	●	●
Color fastness	Washing	●	●	●
	Crocking	◎	●	○
	Perspiration	○	◎	○
	Dry cleaning	●	○	○
	Water	●	●	●
	Ironing	○	●	◎

—continued

TABLE 3.4 (CONTINUED)
Normal Clothing Evaluation

Category	Test Type	Relevancy	Importance	Possibility
	Sea water	○	○	○
Product test	Industrial production	●	◎	●
	Geosynthetics	○	○	○
	Nonwoven filter	○	○	○
	A production of rubber	◎	◎	●
	A button	●	●	●
	A zipper	○	○	●

Note: Standardization priority: high (●), medium (◎), low (○).

3.5.1.3 Test Evaluation and Results

We planned test evaluations according to the evaluation methods of clothing and electric/electrical evaluation methods selected above. The results by the KATRI (Korea Apparel Testing and Research Institute 2008) are shown in Table 3.6.

3.5.2 Establishing a Standard for Performance Evaluation

To establish standardization, we separated the standards into three categories: (1) clothing performance evaluation standard, (2) electricity/electron performance evaluation standard, and (3) physical performance evaluation standard. Therefore, any data acquired from test evaluation is substituted into these three standards, and established as standardization basis. However, considering that this study is in progress, only part of these standardization categories is established, thus the final result will be presented through further study.

3.5.2.1 Clothing Performance Evaluation Standard

It is vital to have a standard for washing performance or size changes, because the fabric signal wire is basically cloth. The textile keypad uses an electric signal and the fabric electrode for sensing living body signals. Therefore, they should be attached to a suitable place inside clothes near human skin. It is also necessary to have safety standards because heavy metals in the clothes come in direct contact with the skin. Therefore, detailed evaluation standards are required. The clothing performance evaluation standards refer to recommendations by KTIKCA (Korea Consumer Agency) (Table 3.7). The standard is directly evaluated and it can be amended and supplemented depending on further testing.

3.5.2.2 Electrical/Electron Performance Evaluation Standard

Smart clothing is basically fabric, but its functions should satisfy safety standards because it includes electrical functions. So, function standards have been established based on reliable evaluations of previous electrical products and stability

TABLE 3.5
Electrical/Electron Product Evaluation

Category	Test Type	Relevancy	Importance	Possibility
Reliability	ISO/IEC 9126-2 8.2.1 Fault rate	●	●	○
	ISO/IEC 9126-2 8.2.1 Problem rate	●	●	○
	ISO/IEC 9126-2 8.2.2 Down rate	●	●	◎
	ISO/IEC 9126-2 8.2.2 Recovery rate	●	●	◎
Safety	Safety of household and similar electrical appliances	◎	●	○
	General performance—safety requirements	●	●	○
	Specification for safety of household and similar electrical appliances.	○	●	○
	Particular requirements for electric irons	●	◎	◎
	Particular requirements for spin extractors	●	●	○
	Specification for safety of household and similar electrical appliances	○	○	○
	Particular requirements for washing machines	●	●	◎
	Particular requirements for shavers, hair clippers, and similar appliances	○	○	○
	Household and similar electrical appliances, safety	●	●	◎
	Particular requirements for floor treatment machines and wet scrubbing machines	◎	●	○
	Audio, video, and similar electricity apparatus—safety requirements	◎	●	○
Electricity	Appliance couplers for household and similar general purposes	●	●	○
	Ballasts for tubular fluorescent lamps	●	◎	●
	Auxiliaries for lamps, a.c. supplied electricity ballasts for tubular fluorescent lamps	●	◎	◎
	Single-capped fluorescent lamps—safety specifications	◎	●	◎
	Information technology equipment	○	○	○
	Safety requirements for electrical equipment for measurement, control, and laboratory use	●	●	○
	Hand-held motor-operated electric tools	●	●	○
	Fixed capacitors for use in electricity equipment	○	○	○
	Testing and measuring equipment/allowed subcontracting	○	○	○

Note: Standardization priority: high (●), medium (◎), low (○).

TABLE 3.6
Test Evaluation and Results

Test Type	Reference Standardization	Methodology	Result
Bending resistance	IEC 62221	Take a moderate length of a sample and hang a 1 kg weight at position 50 cm from the sample's fixed end.	After 30,000 times: normal
Tensile strength	IEC 60811-1	Run about 0.1 A of electric current through the conductor. Tensile speed: 250 mm/min Number of samples: 4	Tensile: 732.5 N/mm^1
Abrasion resistance	KS K 0604	The band must not be broken due to abrasion when the disk is spun 60 times per every minute in the same direction as the gravity of weight.	After 10,000 times: normal
Appearance changes	JIS L 1018 KS K 0815	Clothes form changes after washing Washing times: 15 times	No change
Size changes	KS K 0608	Clothes size change rate after washing Washing times: 15 times Size change rate ±2% below	Size change rate ±1%
Determination of extractable heavy metals in textiles	KS K 0731 ISO 3696	Detect heavy metal after washing Washing times: 15 times Cd, Pb, Hg, Cr, VI, Cr, As, Sb, Cu, Co, Ni, Ba, Se	Not detected
Button lifetime	KS C 4516 IEC 60947-5-1	Cylinder pressure for switch button: 140~300 g • Speed for push: 2 times/s • 1 second on/off action Button is OK after 50,000 times pushes for a test	Malfunction is below 0.1%
R-peak detection rate	IEC 60601-2-27 KS P 1217	An algorithm based on wavelet transforms (WTs) has been developed for detecting ECG characteristic points. By using this method, the detection rate of QRS complexes is above 99.8% for the MIT/BIH database and the P and T waves can also be detected, even with serious base line drift and noise.	Static Above 91.5% Dynamic Avove 73%
Insulation resistance	IEC 60227-1	Test voltage: DC 500 V Test times: 4 times	1 times 4.2×10^3 2 times 3.28×10^9 3 times 1.19×10^6 4 times 1.71×10^8
Circumstance		1. Temperature 24°C humidity 52%.	

TABLE 3.7
Clothing Performance Evaluation Standards

Type	Standard	Numerical Value	Test Condition
Appearance changes	Not changed	±1%	Washing 15 times
Size changes	Size change rate	Below ±2%	Washing 15 times
Heavy metal[a] detection	Not detected	0%	Washing 10 times

[a] Cd, Pb, Hg, Cr, VI, Cr, As, Sb, Cu, Co, Ni, Ba, Se

TABLE 3.8
Electric/Electrical Performance Evaluation Standard

Type	Standard	Numerical Value	Test Condition
Fabric keypad	Button lifetime	Below 0.1%	Testing 50,000 times
Fabric transfer wire	Insulation resistance	Above 10^8 Ω/cm	DC 500 V
	Electrical resistance	Below 40 Ω/m	DC 500 V
Fabric ECG electrode	R-peak detection rate	Static: Above 90%	Testing 1000 times
		Dynamic: Above 70%	

certifications of those electrical appliances. The standard for the fabric keypad refers to "RS B 0177" for an electron switch or button. The fabric signal wire refers to "KS C IEC 60885-2" for electrical properties. In the case of the ECG electrode, it refers to "KS C IEC 60601" (Table 3.8).

3.5.2.3 Physical Performance Evaluation Standard

The evaluation methodology for tensile strength is found in textile areas and electrical product areas. Following research, a tensile strength test should distinguish between normal clothes and electricity, depending on the requirements. For fabric conductors that contain wires inside, the check must adhere to the KETI (Korea Electric Testing Institute) wire evaluation standard.

But signal wire as a textile substitute material is not affected by physical impact as much as electrical wires or external environments. The Martindale method is used for textile abrasion resistance. The standard of that tearing strength is 10,000 times, followed by KCA (Korea Consumer Agency) recommendation (Jo and Lee 2003).

The bending resistance test refers to an electric/electron test or "bending test," as there are no specifications in textiles areas. The standard for the flexural test was set to 20,000 times, as recommend by an expert at KETI.

If a fabric keypad is attached on the clothes, the seam strength test, which includes seam areas on the clothes, will gradually increase the tensile strength machine until the fabric is torn.

Evaluation items and their standards will be amended and fixed by continuous test analysis (Table 3.9).

TABLE 3.9
Physical Performance Evaluation Standard

Type	Standard	Numerical Value	Test Condition
Fabric transfer wire	Abrasion resistance	No malfunction	Testing 10,000 times
	Tensile resistance	Above 500 N/mm^2	Tensile speed: 250 mm/min
Fabric keypad	Bending resistance	No malfunction	Testing 20,000 times
Fabric ECG electrode	Bending resistance	No malfunction	Testing 20,000 times

3.5.3 FURTHER SCHEDULE FOR EVALUATION

The evaluations are divided into three areas: (1) clothes, (2) electric/electricity, and (3) physical. Subsidiary materials will also be selected and applied to three evaluation areas depending on their properties.

Currently, all clothing evaluation has been established except for stitch strength. However, a test for stitch strength will be established in the future. Electrical/electron testing has been completed for insulation resistance and electrical resistance. The test for button reliability and sensitivity is in development.

Regarding textiles, the deformation of appearance and size change after washing was tested by the Korea Consumer Agency's recommended quality standard. The post evaluation is being carried out now.

3.6 STANDARDIZATION STRATEGY FOR SMART CLOTHING

3.6.1 ANALYSIS FOR TECHNOLOGICAL DEVELOPMENT

In order to establish standards, many investigations into related research and product commercialization should be accomplished at the same time. Therefore, technological developments regarding standard level, technology lifetime, and marketability are required for this standardization. Technology level means the possibility and value for development, and technology lifetime is divided into three phases, which are "developing," "before developing," and "testing." In short, a higher standardization level has more influence on the technology development. The shaded parts of Table 3.10 are especially regarded as having a high tactical value of technology development.

3.6.2 FURTHER RESEARCH ON RELATED TECHNOLOGIES

We obtained critical information about the present level of standardization and technological development, which is expected to be utilized for future standardization research through standardization analysis for smart wear. To establish standardization strategies, however, we need to open up a various and detailed field. Therefore, in order to expand the study scope of smart clothing technology development and standardization, we selected wearable computing and U-healthcare technology. The following is the overall review of these two fields. If we select technology and standardization factors that can integrate with smart clothing, we expect to expand smart clothing standardization study.

3.6.2.1 Wearable Computing

3.6.2.1.1 Definition of Wearable Computing

Wearable computing refers to a computer that is inserted into the personal space of the user, is controlled by the user, and has both operational and interactional constancy, i.e., is always on and always accessible. Most notably, it is a device that is always with the user, and into which the user can always enter commands and execute a set of such entered commands, and the user can do so while walking around or doing other activities (Mann 1998).

3.6.2.1.2 Wearable Computing Development Areas

- *Military area*: Studies are under way to develop wearable computing for military purposes, as it was first introduced about 30 years ago. There are many kinds of wearable computing, like helmet with display, light sensor, sounder and backpack-computer, GPS, various guards, and military uniforms. Investments in wearable computing in the military market are expanding their scales with predictions that there will be information warfare in the future.
- *Industry area*: Wearable computing has been used for offering operation orders and user manuals, and for sensing the malfunction of products, for 10 years. The Boeing Company, as well as FedEx and NASA use wearable computing by loading a plan and required technology to the clothes required for work in the various environments. Wearable computing has been expanded to offer not only convenience and efficiency, but also technical knowledge for users.
- *Medical area*: Smart shirts and clothing have been made by several companies including Sensatex (Sensatex 2007). Their shirt contains sensors that can be used to monitor vital signs such as heart rate and pulse and transmit the data to a medical center. It also provides a sense of security for patients and automatically calls an ambulance in emergencies. The Starlab's (Starlab 2008) attempt to build an artificial brain in Belgium is under way to study intelligent clothing products as substitutes for computers, cell phones, and monitoring equipments. Lifecore Inc. (Dickinson 2002) has commercialized a defibrillator that is programmed to detect abnormal heart rhythms and correct them by delivering a jolt of electricity to patients with cardiac disorders. These are now expected to be developed into products with self-treatment abilities and will be a great help to hospital staffs.
- *Entertainment/fashion area*: Music players and ICD+ jackets containing devices for music and video interfacing are becoming the most popular products in this area. More diverse applications are going to be loaded to established products with MP3 video capabilities.

TABLE 3.10
Analysis for Technology Development

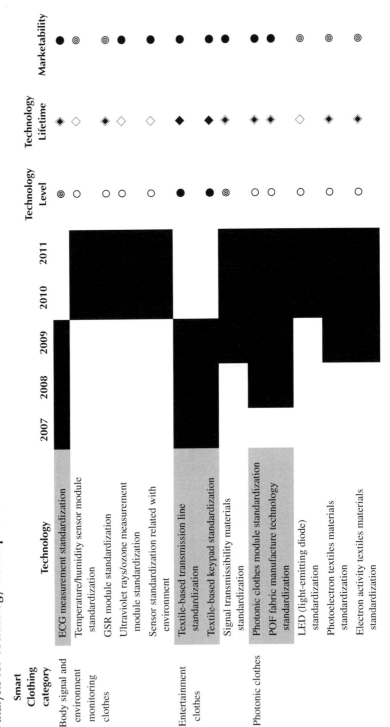

Smart Clothing category	Technology	2007	2008	2009	2010	2011	Technology Level	Technology Lifetime	Marketability
Body signal and environment monitoring clothes	ECG measurement standardization						◉	◆	●
	Temperature/humidity sensor module standardization						○	◇	◉
	GSR module standardization						○	◆	◉
	Ultraviolet rays/ozone measurement module standardization						○	◇	●
	Sensor standardization related with environment						○	◇	●
Entertainment clothes	Textile-based transmission line standardization						●	◆	●
	Textile-based keypad standardization						●	◆	●●
	Signal transmissibility materials standardization						◉	◆	
Photonic clothes	Photonic clothes module standardization						○	◆	●●
	POF fabric manufacture technology standardization						○	◆	
	LED (light-emitting diode) standardization						○	◇	◉
	Photoelectron textiles materials standardization						○	◆	◉
	Electron activity textiles materials standardization						○	◆	◉

Extra functional clothes

Smart clothing interface standardization

Smart clothing components standardization

Smart clothing value evaluation standardization

Smart clothing terminology standardization

Note: Standardization priority: High (●), Medium (◎), Low (○), Developing(◆), Before developing(◇), testing (◆).

3.6.2.2   U-Healthcare

3.6.2.2.1   Definition of U-Healthcare

Traditional healthcare is undergoing a paradigm shift to "ubiquitous healthcare," especially with the advent of home networks. Many nations have already focused on tele-medical service with U-city concepts. For instance, a U-healthcare business project is actively in progress in the European Union (MobiHealth), the United States (LifeShirt and LifeGuard), Japan (SELF: Self Environment for Life), and many other countries (Electronics and Telecommunications Trend Research Institute 2006). U-healthcare means that the safety assurance of a patient and quality of the medical service used for the health management of disorders does not have to be abandoned in order to provide a service (Lee 2006). It is the IT-based medical service that accumulates and regularly checks individual medical information and transmits the data to a medical center, making it possible to have home medical service for prevention, diagnosis, treatment, and aftercare. The elderly living alone, or patients with chronic disease like diabetes, obesity, and high blood pressure are the main customers of these services.

3.6.2.2.2   Importance of U-Healthcare

Many countries in the world are headed towards becoming aging societies due to increased average life span and declining birthrates. Clearly, this kind of society that is demographically aging is a critical condition, but also benefits the world's economies. The importance of U-healthcare is greater than ever. Companies working for IT such as Intel Co. and IBM (International Business Machines) have already branched out into U-healthcare or are extending their business abroad. U-healthcare, which has potential as a new growth industry, can be a new medium for curtailing medical expenses in Korea's economy. It is estimated that more than $1.5 billion Wons in National Health Insurance can be cut down only by remote monitoring systems for patients (Kang 2007). Also, it can be the way to lower irrational overheads incurred by medical mistakes, positioning itself as one of the regional leaders for the future ubiquitous world. Smart clothes with a monitoring system for the body will be a part of U-healthcare, which is significantly important for the future.

3.6.3   Expected Effects of Standardization

3.6.3.1   Revitalization of Fashion Industry through Standardization of Technology

As a result of new industry appearance like 6T, which was called IT (information technology), BT (biology technology), NT (nano technology), ET (environment technology), ST (space technology), and CT (culture technology) convergence, fashion industries are changing their direction from labor-intensive to knowledge-intensive industries. The fashion industry not only makes clothes, but also progresses into the knowledge industry, which blends technology, culture, and image. And such a knowledge-intensive, high-valued textile industry is expanding rapidly. Therefore, the fashion industry will be in a strong position in the world market. The world is united into one global economy system with rapid change. Standardization of core technology stands out as an important issue. Standards are necessary for the industry's development and competition in the global market where securing global standards is fierce. At the same time, it will be able to revitalize the clothing industry's parallel with developing technologies as well as participating in the standardization.

3.6.3.2 International Competitiveness through Standardization

The standardization of specifications, telecommunication, protocol, and systems for a product's function can encourage competitiveness in the international market, easy entry to the market, and leadership in the industry.

3.6.3.3 Increase Market Sharing with Technology Development in Parallel with Standardization

It is clear that technology development is required for increasing market share. Therefore, market sharing for smart clothing in the international market can be maximized by standardization along with technology and product development.

3.6.3.4 Positioning in International Standardization Society

If this study dominates the technology area, such as digital textiles and e-textiles, before international specification has been ratified, it will be able to take the lead in the international smart clothing market. The responsibility for standardization of the technology can reside in a chairman, director, or other high-ranking official of the ISO/IEC. Therefore, our profile will be high and we will have a step forward when competing with the other developed countries' monopolies and the challenges posed by emerging markets.

3.7   CONCLUSION

Smart clothing is the convergence of the textile/clothing and digital industries. As the combination between these two industries brings large opportunities and benefit, many institutions, organizations, and enterprises have paid great attention and planned significant investments. However, the standardization of smart clothing technology has rarely been considered and little research has focused on solving these problems. Nevertheless, this study describes the current situation of standardization in the smart clothing area and presents how strategic thinking may help toward achieving standardization of smart clothing. As a result, standardization strategies can be used for gaining the advantage in the smart-wear market and raising its efficiency. In addition, it can be expected to occupy a more competitive position for standardization. Hereafter, this study will verify feasibility and objectivity of the performance evaluation standard and its method. For this, this study will classify various forms of smart clothing according to a technology road maps and execute test evaluation. Moreover, this study will present a technology road map for a global standard through constant interest in foreign technology and standardization trends related to smart clothing.

REFERENCES

Ahn, Y. M. 2004. Ubiquitous computing clothes. *Textile Technology and Industry* 8 (1): 1–10.

American Society for Testing Materials (ASTM). 2008.

BBC Research. 2007. Smart and interactive textiles. http://www.bccresearch.com/report/ AVM050B.html.

British Standards (BS). 2008. http://www.standardsuk.com/.

Busayawan, A. 2003. The future design direction of smart clothing development. http://bura .brunel.ac.uk/bitstream/2438/1362/1/Textile+institute+journal.pdf.

Caferi, M. 2007. Kuchofuku air-conditioned bed and clothing. CScout Japan. http://www .cscoutjapan.com/en/index.php/kuchofuku-air-conditioned-bed-and-clothing/.

Cho, G. S. 2006. *Latest clothing material.* Seoul, Korea: Sigma Press.

Cho, G. S. 2007. Yonsei University, Institute of Smart Clothing Development. *Korea Society for Clothing Industry Journal* 9 (4): 455–56

Comprehensive Merchandising Support. 2007. Eleksen a peratech company. http://www.elek-sen.com/?page=news/index.asp&newsID=78.

Deutsche Industrie Normen (DIN). 2008. http://www.din.de/cmd?level=tpl-home&language id=en.

Dickinson, J. G. 2002. Medical device and diagnostic Industrial report. (Lifecore Inc.). http:// www.devicelink.com/mddi/archive/02/03/009.html#1.

Electronics and Telecommunications Research Institute (ETRI). 2007. Electronics and tele-communications trend analysis. 22 (1246): 22–23.

Harmon, A. 2007. Eleksen bringing iApparel to mass market. Defining Men's Fashion. http:// www.dnrnews.com/site/article.php?id=552.

International Organization for Standardization (ISO). 2008. http://www.iso.org/iso/home.htm.

Ishimaru, S., K. Chowa, and E. Kogaku. 2006. Microclimate within clothing based on COOL BIZ. 80 (7): 511–14.

Japan Industrial Standards (JIS). 2008. http://www.jisc.go.jp/eng/.

Jo, H. K, and S. H. Lee. 2003. Reversion of the quality standards recommended fiber products. KCA's recommendation (Korea Consumer Agency). http://www.kca.go.kr/front/infor-mation/inf_01_08_view.jsp?no=1122&ctx=0105.

Kang, S. U. 2007. The advent of U-healthcare's age: Samsung economic research institute, 2007 Korea Apparel Testing and Research Institute. 2008. Quality standard informa-tion. http://www.cleaningq.co.kr/quality.php.

Korea Apparel Testing and Research Institute (KATRI). 2008. International environmentally regu-lated toxic substance. http://www.katri.re.kr/board/board_view.asp?boardid=4&num=49.

———. 2008. http://www.katri.re.kr/customer/guide_03.asp.

Korea Consumer Agency. 2003. Summary of amendment on recommended quality standard of textile products. Seoul, Korea.

Korea Federation of Textile Industries (KOFOTI). 2008. A criterion of general clothing per-formance. www.kofoti.or.kr.

Korea Industrial Technology Foundation. 2007. Road map for textile industry technology. http://www.kotef.or.kr/pds/tech_view.asp?RecordID=185&AccessFlag=&cCode=4&p age=1&field=&key.

Korea Patent Information Institute. 2006. Smart textile technology and market. Patent analysis report. Seoul, Korea.

Korean Standards (KS). 2008. http://www.kats.go.kr/.

Korea Textile Development Institute Textile Information Team. 2008. Product development and innovation. Seoul, Korea.

Lee, H. S. 2006. IBM U-HealthCare: Tutorial of Database Technology for U-Healthcare Bio Medical Industry.

Mann, S. 1998. Wearable computing as means for personal empowerment: Keynote address for the first international conference on wearable computing. http://wearcam.org/wearcompdef.html.

Marzano S. 2000. *The quest for power, comfort and freedom. New nomads: An exploration of wearable electronics by Philips*. Rotterdam: 010 Publishers, 4–9.

McWilliams, A. 2007. BBC research. Smart and interactive textiles report. http://www.bccre-search.com/report/AVM050B.html.

O'Neill. 2008. Comm. Emt Jacket. http://www.oneilleurope.com/h3/H3_Manual.pdf.

Orth, M., R. Post, and E. Cooper. 1998. Fabric computing interfaces. http://web.media.mit.edu/~morth/home.html (accessed July 26, 2002). Also appeared in *Proceedings of Conference on Human Factors in Computing Systems*. Los Angeles: ACM Press.

Park, W. 2007. Zegna Sport announces iJACKET—The Bluetooth jacket with touch-sensitive smart fabric. Into Mobile. http://www.intomobile.com/2007/08/22/zegna-sport-announces-ijacket-the-bluetooth-jacket-with-touch-sensitive-smart-fabric.html.

Reid, D. 2006. Smart fabrics are back in fashion. BBC NEWS. http://news.bbc.co.uk/2/hi/programmes/click_online/5286594.stm.

Sensatex. 2007. Smart Shirt. http://www.sensatex.com/smartshirt.html.

Son, B. M. 2007. Fiberwear digital. EBN industrial news fabric. http://www.ebn.co.kr/news/n_view.html?id=294447.

Starlab. 2008. The study of intelligent clothing products. http://starlab.vub.ac.be/website/.

Van Heerden, C., J. Mama, and D. Eves. 2000. *Wearable electronics. New nomads: An exploration of wearable electronics by Philips*. Rotterdam: 010 Publishers, 14–22.

Wearable Computer Lab. 2008. http://wearables.unisa.edu.au/.

4 Electro-Textile Interfaces
Textile-Based Sensors and Actuators

Kee Sam Jeong and Sun K. Yoo

CONTENTS

4.1   INTRODUCTION

Just as the history of a costume reflects the changing culture and technology of its age, the purpose of clothing has also been changing with the culture of its age. The purpose of clothing in the primitive age was to protect the body from the environment. The animal leather and furs and the leaves of grass not only protected humans from nature but also extended their living space to include the severe cold area where human activities were difficult without the protective clothing. The function of clothing has evolved from the means of protecting human beings to the instrument of augmenting human capabilities. The biblical phrase in Genesis 3:7 expresses the clothing's socio-cultural and communicational functions quite clearly: "And the eyes of them both were opened, and they knew that they were naked; and they sewed fig leaves together, and made themselves aprons." The clothes after the Middle Ages had placed emphasis on beauty and authority, and after the 20th century, not only fashion but also functionality was emphasized.

Then what kind of role is clothing supposed to play in the knowledge age of 21st century? The latest development in information and communication technology (ICT) has changed the conventional lifestyle completely, and the role of clothing too is expected to go through radical changes in the rapidly changing environment. Smart clothing is at the core of these changes. The communications theorist Marshall McLuhan made the following comment in his work in 1968: "The computer is the most extraordinary part of Man's technological clothing: it is an extension of our central nervous system" (McLuhan 1968).

This may be the very first statement that may help define the concept of smart clothing. Viewed from the perspective of the modern age, smart clothing may be defined as the apparels that combine the functional materials and electronic technologies to enhance humans' adaptability to the environment. Many researchers interpreted "smart or intelligent" to mean the ability to perceive the surrounding environment and react to it (Baurley 2004). This is comparable to the ubiquitous computing ability of "all things to communicate and react with other things to provide services to human beings." As shown in Figure 4.1, in the ubiquitous age, clothing in the personal area will play the role of a gateway connecting humans with the environment. Clothing of the future will be linked to the information and communications infrastructure and provide quiet service to humans by perceiving the environment and at the same time by providing the information generated in personal areas such as health, emotions, and locations of person to the environment. To carry out these functions, all the information systems are equipped with input and output functions. In the smart clothing, the sensor and the actuator are responsible for these functions. From this point of view, Zhang and Tao have classified the extent of intelligence

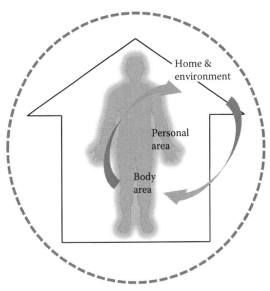

FIGURE 4.1   Media objects in the future living environment.

into three categories (Zhang and Tao 2001; Van Langenhove and Hertleer 2004). *Passive smart textiles* refer to the ability of the general sensor that can perceive the environment. *Active smart textiles* not only perceive the environment but also react to internal and external stimulation. They have the sensor function as well as the actuator function. *Very smart textiles* are several steps more advanced than the *active smart textiles*, possessing the ability to adapt to changing situations. In addition to the sensor and the actuator, the *very smart textiles* have a microprocessor in which an intelligent agent is installed in order that they may provide appropriate services according to the changing situations. The *very smart textiles* are similar to the concept of the smart object in ubiquitous computing (Weiser 1991).

What is in direct contact with or closest to the body in the ubiquitous space is the clothing. The apparel is the border that distinguishes the body area from the environment. Smart clothing senses the information about its wearer's body and informs the wearer of the conditions of the body or sends the information to the outside world, while at the same time it informs the wearer of the information or services available in the external world. Needless to say, smart clothes can also just store that information. To carry out these functions, smart or intelligent textiles must possess special properties that the conventional fiber does not have. The clothing must have a sensing function in order for it to perceive such variables as biomedical signals and body temperature of its wearer. Furthermore, it must also have the actuator function to inform its wearer of the information or services available in the external world. It would be an ideal case if the fiber itself becomes the sensor or has a built-in actuator function. However, that is not yet possible, although there is growing potential (Linz 2007) with the leading edge advances in materials science, fiber engineering, and nanotechnology. In this chapter, we discuss the products that are applicable and available for smart clothing and the way in which the fiber can be made intelligent by applying electronic technology.

Electro-textile interface is indispensable in producing smart clothing, but it is very difficult to integrate the electronic products and fabric products because their manufacturing processes and their physical properties differ greatly. The factors that make the electro-textile interface difficult are discussed below.

First, fabric is very flexible, whereas most electronic components are solid. Parts of electronic products are made of inorganic materials or metals including semiconductors. Therefore, most materials and components are either solid or encased in solid housings. Since most textiles are made from short fibers or long fibers that have been twisted, they are very thin and flexible and their shape can be changed and they can even be stretched when force is exerted. When electronic components and fabrics are integrated, stress is focused on the boundary between the soft and solid parts, and this causes a negative influence on the reliability of the product. Moreover, the solid parts in clothing markedly reduce the satisfactory wear feeling, whereas the flexibility of fiber itself is a factor that goes against the precision or manufacturability of interfacing.

The second major difference is electric conductivity. Electronic parts have the characteristics of conductor or semiconductor and are activated by electric current flowing from the power source. However, the majority of fiber is non-conductive by nature. Although some new products have recently been introduced that use metal yarn or conductive polymer, because of lack of conductivity and flexibility their use is extremely limited.

The third difference is air tightness and water resistivity. The material of most electronic apparatus itself does not absorb liquid, but the electrical system could be damaged by water once it is integrated to printed circuit board (PCB). On the other hand, clothing is constantly in contact with liquid like water because it not only has to absorb and discharge sweat but also has to be washed. Therefore, special measures have to be taken to make it waterproof.

In addition to these differences in physical properties, the textile industry and the electronics industry have problems with technical terms, industrial standards, and manufacturing processes between the two industries. The textile industry is using such units as yard, pound, and denier, but the electronics industry is using units such as meters, grams, voltage, and ampere. The differences in standards will not only create difficulties in research for researchers involved in both fields, but also become serious obstacles hindering the effort for mass production through industrialization. There is a cultural barrier also, because specialists of each industry seek different goals. We are urgently in need of specialists who can break out of the stereotypical way of thinking and connect these two vastly isolated fields.

The differences in manufacturing processes are a very difficult problem to overcome. Electronic parts are mounted and soldered on PCB, but clothing is manufactured through a great variety of weaving, post-processing, and sewing techniques. Most of the fabric products cannot stand over 200°C of heat, so soldering cannot be applied on them. Therefore, we must develop new materials as well as new engineering processes that would fuse the two industries seamlessly. A researcher involved in the study of smart clothing is confronted with numerous problems besides the ones mentioned thus far. In this chapter, we examine the ways in which fiber and electronic devices can be integrated.

4.2 ELECTRO-TEXTILE INTERFACES

The emergence of mobile devices and miniaturized electronic apparatus has had a great deal of influence on modern people's life patterns. Many people can enjoy music during jogging. Mobile phone technology allows people to answer phone calls anywhere. As many people came to carry many portable devices on themselves, their clothing too came under the influence of changing lifestyle. The early smart clothing provided space in the clothes needed to carry portable devices conveniently as well as wire-guides (or path) for earphones. This method is generally applied to the latest outdoor clothes or clothes for juniors. In the next stage that followed, electronic devices are integrated to clothing. Fiber-based electric wires that use metal fiber are inserted into the clothes, and electronic devices such as MP3 or keypad are attached to the clothes (Weber et al. 2003). Many of these products are gaining popularity recently in the outdoor wear market.

However, these products do not mean that electronics technology and textile technology have been completely integrated. They only mean that the products manufactured by both industries are utilized in simple combinations. Smart clothing literally means that fiber should be a sensor or actuator in itself and perform the functions that the user desires, and that the clothing and the environment should be able to communicate interactively and provide invisible services to the people. To realize this goal, textile must have various active functions in itself and electronic products should be fused into the fabric. There are few materials now that have electronic functionality that currently available technologies can offer. To make smart clothing, we need to develop a new technology that can integrate and fuse electronic products with fiber. That new technology is called electro-textile interface.

4.2.1 CONNECTION BETWEEN OBJECTS

With the technology available now, we have no other choice but to use electronic devices to implement active functions in smart clothing. There are some smart materials that can react to a stimulus like thermocolor (chameleon) paint. Color of the material varies passively at a given temperature, but we cannot control color actively. All electronic products carry out their functions by controlling the electric current flowing into objects that have independent functions. A variety of objects can be used in smart clothing such as a special functional material, a resistor, an integrated circuit (IC), a PCB on which many electronic components are integrated, or a device with an independent function. Whatever form and shape the objects may take, if there are some electronic objects in the clothes, they will need the physical current path to connect them and to supply electric power. For example, when a textile-based keypad is placed on the sleeve and an MP3 is positioned in the inside pocket, signal transmission lines will be necessary. If an extra power source is required for an independent object to perform an active function, power lines are needed. There are wired and wireless methods of transmitting signals. Wireless signal transmission method has many problems, such as power consumption or electromagnetic interference (EMI) to other devices and the human body, to be applied in the clothing environment. The biggest problem is how to stably supply the power to the wireless system. Although wireless signal transmission is convenient, it consumes much more electric power than wired signal transmission. Unlike portable electronic products, clothes are worn constantly. In other words, clothing is providing its service 24 hours for its wearer. Smart clothing cannot ignore this essential feature of ordinary clothes. If conventional clothing provided passive service, the clothing in the new age should have an intelligent agent installed in it and provide voluntary calm service for 24 hours. A larger and heavier battery is required for long hours of use, and this not only adds to the user's inconvenience but also imposes various limitations on clothes designing. Consequently, power supply, especially the battery, is the first problem to be considered not only in designing portable apparatus but also designing smart clothing. One of the solutions for this power problem is to apply radio-frequency identification (RFID) technology.

RFID component is supplied with energy from an outside device when an event is triggered. RFID technology is expected to play an active role in smart clothing. Another problem is the issue of radio wave interference and security. Already many people are carrying various radio wave sources such as cellular phones on themselves. Since there are a great variety of noises in the propagation paths of radio waves and the radio bands we are allowed to use are limited by law, it is imperative to always consider the radio wave interference problems when we design wireless equipment. Security is another problem because the wave path of radio waves is impossible to control by nature. Still another problem is the psychological fear of EMI. Just as the EMI from cellular phones has turned into a social issue (Food and Drug Administration [FDA] 2003), the EMI from smart clothing must be dealt with even if wireless equipment is not used.

Wired signal transmission too has many problems for smart clothing. The polyvinyl chloride (PVC)-coated copper wire that is generally used for electric wire in the electronics industry is too stiff and too thick to be used for smart clothing. Moreover, we should be careful about using copper because it is easily oxidized by sweat or water and that could generate chemicals hazardous to the human body. An ideal wire to be applied in the clothing environment would be a material that is chemically stable and at the same time possesses a physical property that is identical to textile. The most ideal material for smart clothing would be a polymer fiber that has a copper-like conductivity. To date, such a material has not yet been developed. The currently available materials are limited to thread made from metal yarn, thread coated with metal powder, and metal-plated material. These materials are handicapped as electric wire because their conductivity is insufficient and they are weak in washing and in mechanical changes of fabric structure.

4.2.2 Electro-Textile Platform

For the electronic objects to co-exist with textile in the clothing environment despite the numerous difficulties mentioned above, and to implement the common functions, they need an electro-textile platform, which is defined as the infrastructure to be shared with electronics and textile. The electro-textile platform can be classified into two categories in accordance with the application area: *micro-platform* for interconnection between components in a relatively small area, and *macro-platform* for interlinking with each subsystem in the entire clothing (Figure 4.2).

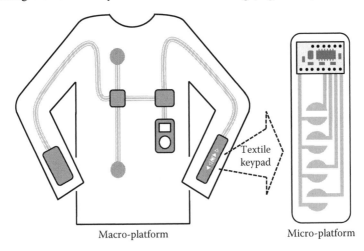

Macro-platform   Micro-platform

FIGURE 4.2   Electro-textile platforms.

The *micro-platform* plays the role of a PCB in the electronic product, which connects parts to other parts to implement a sub-system. Most of the *micro-platforms* can be put together on a single sewing pattern piece. A *macro-platform* connects one object to another or various objects spread apart within clothing and make up an integrated system. In general, a *macro-platform* goes over the boundary between the patterns. A *micro-platform* does not affect the designing of clothing since it occupies a small portion of the local area. On the other hand, a *macro-platform* gives an influence on the entire clothes manufacturing process ranging from the selection of material to designing, sewing, and post-processing. This can be a very difficult task.

To meet the requirements of design and functions with heterogeneous materials is a great and complicated challenge for which clothing designers, pattern designers, merchandisers, sewers, electronic engineers, and marketing specialists would have to agonize over a number of hard problems. Commonly used materials and methods for construction platforms are shown in Figure 4.3.

Use of narrow bands is the most common method. The narrow band is similar to the flat cable in the electronics industry, and it is woven with several threads of conductive fiber placed in parallel. In general, metal yarn or thread coated with metallic particles (usually silver-coated thread) is used as conductor. To prevent possible short circuits in between wires, threads are generally coated with insulating material.

Sewing and embroidery are also one of the readily available methods. Conductive threads are often sewed or embroidered along the circuit paths. This method too generally uses metal yarn or threads coated with metallic particles. Metal yarn is stiffer than ordinary thread and causes difficulties in embroidering for mass production. In some cases, specially designed embroidery machines are required because ordinary embroidery machines are not good enough. It is easier to embroider the metal-coated threads than metal yarn, but their drawback is the loss of electric conductivity when washed. As in the case of the narrow band, short circuits in between wires are another weakness of the embroidery method.

Conductive ink printing composes the circuit in a method similar to the PCB manufacturing process, which can be theoretically the simplest and easiest way. It is also easy to make the mass production system. The circuit can be made by silk-screening or by using a digital textile printer (DTP). Commercial use of the printing technology is faced with the following problems. The biggest problem is the procurement of highly conductive ink. Electrical resistance is proportional to wire length and inversely proportional to cross sectional area. Printing technique by its nature forms a very thin layer of ink, which means that the cross section of the wire necessarily becomes extremely small. Therefore, it follows that the ink layer to be printed should be made thick and highly conductive at the same time.

Polyaniline and polypropylene are conductive polymers that can be used for smart clothing, but they do not have good conductivity. Conductive ink can be made by mixing conductive carbon blacks or metal nano-particles, but it is also difficult to get sufficient electrical conductivity. If the amount of carbon powder or metal powder is increased to enhance the electrical conductivity, the ability to bind to textile is lowered. Another drawback to carbon powder or metal powder is its weakness to physical changes in shape and form. Repeated physical changes in shape create cracks on the surface, and the cracks cut off the electric current, disabling its functions.

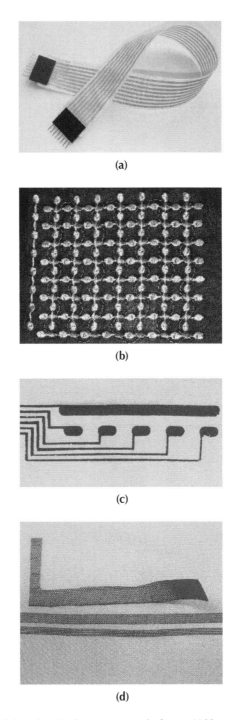

FIGURE 4.3 Materials and methods to construct platforms. (a) Narrow band; (b) Sewing and embroidery; (c) Conductive ink printing; (d) Bonding two or more heterogeneous textiles.

The greater the elasticity of fabric, the greater the chances of cracks on the surface. It is very important to find an optimum combination of the binder and conductive materials that can endure physical changes in shape and at the same time satisfy the conductivity required in smart clothing.

Bonding is a technique that enhances the functionality by bonding two or more heterogeneous textiles. An excellent electro-textile platform can be made by cutting the electroless metal-plated fabric into the pattern of a desired circuit and bonding it to ordinary textile. Since metal is directly coated on the fiber, the electroless metal-plated fabric has electric conductivity superior to any other materials. Its weakness to washing can be overcome by coating or covering methods. The main problem is precision cutting of the patterns is difficult, so laser cutting equipment is required. The bonding technique may be applied to small-scale manufacturing. Besides the methods that have been introduced here, there can be a great variety of methods. Perhaps the fastest and easiest way to find them would be for specialists in various fields to get together and brainstorm for new ideas.

4.3 TEXTILE-BASED SENSORS

As long as a physical change is observed against external stimulation, any material can be used as a sensor. Since the structure and material of textiles differ from semiconductors or inorganic objects that are used in the electronics industry, those who are designing textile sensors need a radical shift in their conceptualization. Let us examine a few characteristic features of textile before designing the textile sensor.

Textile is composed of fiber. There is woven fabric and non-woven fabric. Woven fabric has a variety of structures depending on the method of weaving: plain weave, twill weave, satin weave, knitted fabric, etc. After it is woven, the fabric goes through various processes such as dyeing, softening, antistatic finishing, etc. The many kinds of textile thus created can be used as the sensor according to their mechanical and chemical properties. If a certain change occurs in the woven fabric, that change must be recorded into the analog/digital processing device. For this purpose, the change has to be converted into electric quantity. And then, to transmit converted electric signal, conductive fabric is required. The material that is used for the sensor must have sufficient conductivity to transmit the output of the sensor to the system. An ideal textile sensor would be the fabric that can generate electricity or change the flow of electric current in its reaction to external stimulation. Unfortunately, however, with the currently available technology, it is almost impossible to obtain woven fabrics or compounds possessing such functions. It follows, therefore, that among other alternatives, we must also consider combining electronic components with fabric material to the extent that would not hinder the user's activities.

4.3.1 MEASURANDS AND SENSORS

The sensor is a device that converts the object that we desire to measure or external stimulation into a physical quantity (electric signals in most cases) that we can handle. Although in most digital systems of our age, physical or chemical quantities are converted into electric signals, not all the sensors are necessarily outputting electric signals. Chemical sensors, for example, convert changes in color, volume, or temperature into electric signals through secondary processing in the digital system.

The sensor can be classified into physical, chemical, electrical, and biological sensors. Then what would be the objects of sensing in smart clothing? They can be divided into two large categories. One is biomedical signals, including human gesture, and the other is environmental variants. Since clothes are the objects that stay closest to the human body 24 hours a day, they are the best platform to consistently measure biomedical signals without bothering wearers. Figure 4.4 shows the representative types of signals that can measured from the human body and the positions at which they can be measured.

Signals that are relatively easy to measure and most frequently used are body temperature, respiration, pulse, electrocardiogram (ECG), electromyogram (EMG), electroencephalogram (EEG), galvanic skin response (GSR), gesture, etc. In general, ECG, EMG, EEG, and GSR use the electrodes for measuring (Burke and Gleeson 2000; Dias et al. 2005; Jang et al. 2007; Karilainen, Hansen, and Müller 2005; Lee, Pearce, and Hib 2004; Sung, Baik, et al. 2007a; Sung, Yang, et al. 2007b). Strictly speaking, the sensor should be distinguished from the electrode, but in this chapter they are described under the same category. Although not included in the category

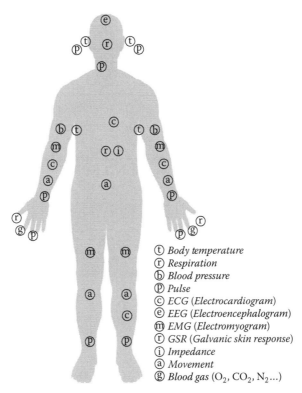

FIGURE 4.4   Biomedical signals that can be measured from the human body.

of biomedical signals, the input mechanism such as textile keypad for man-machine interface also utilizes the sensor technology.

The sensors measuring the environment include those of gas, temperature, moisture, ultraviolet radiation, light, acoustics, etc. They include what is called context awareness function in ubiquitous computing. Furthermore, various other sensors can be applied for the construction of a smart clothing system. Table 4.1 is a list of sensors that are used for the clothing environment among those used in the conventional electronics industry.

The body temperature that is measured at a certain situation can become a piece of useful information by itself. The body temperature measured continuously can offer more information. The human body has a daily cycle called circadian rhythm, and the thermoregulation system tries to maintain its homeostasis, which is disturbed by the changes of circumstances or emotion. Thus, temperature changes with time are useful in understanding the conditions of the wearer. Moreover, it is also possible to extract information that is entirely different from the measured information through secondary manipulation.

For example, emotional information can be extracted by using the ECG (McCraty et al. 1995). The original purpose of the ECG is to diagnose cardiac diseases. ECG signals are weak signals of about 1 mV, and they are liable to be exposed to various noises in the measuring process. ECG signals measured in the clothing environment are expected to be very poor in quality compared to those measured in the hospital. These facts raise some fundamental doubt about the use of ECG signals in smart clothing. In general, the morphology test and the pacing test are the two methods used in the hospital for ECG diagnosis. The morphology test is possible only when signal quality meets criteria of medical equipments, but the pacing test is possible even when the signal quality is relatively low.

In the sports field, heart beat information is extracted from the ECG, and the amount of physical exercise is calculated on the basis of heart beat, and then the types of exercise are prescribed. It is possible to assess the function of the autonomic nervous system (ANS) by applying the heart rate variability (HRV) technique and measure the changes in sensibility. It is important that smart clothing could warn wearers whether they have to see a doctor or not. But, it is very difficult to tell them what kind of disease they have. So, information on the amount of physical exercise or on emotional changes is more useful in the smart clothing field rather than diagnosis of cardiac diseases. This means that the objects to measure and the purpose of sensing in smart clothing should be clearly distinguished from those in the medical or electronics fields. Therefore, researchers involved in the field of smart clothing are obliged to decide what kind of information they are seeking first, even if it may be costly to select the sensors that are most suitable to the clothing environment and the sensors that make the wearers most comfortable.

4.3.2 Textile Electrodes

ECG, EEG, EMG, and GSR are the most frequently used biomedical signals. These signals are measured through electrodes. In hospitals, disposable electrodes are used to extract these biomedical signals. Coupling gel is applied between the electrode and

TABLE 4.1

A List of Sensors That Can Be Used for the Clothing Environment

Measurand	Sensor	Remark
Body and environment temperature	Thermocouple	Output: difference of potential
	Thermistors	Output: difference of resistance
Respiration	Strain gauge	Pressure measurement by chest volume change
	Electrode	Impedance measurement by chest volume change
	Thermocouple	Difference of temperature during inspiration and expiration
ECG, EEG, EMG, GSR, EOG …	Electrode	To measure bio-potential difference or impedance between electrodes
Movement	Accelerometer	2-axis and 3-axis sensors
Gesture	Strain gauge and accelerometer	Movement and posture measurement
	Optical fiber	Measuring changes of transmissivity by optical fiber form transformation
Pulse	Strain gauge	Pulse pressure measurement on body surface
	(Infrared) LED sensor	Volumetric measurement of blood flow in peripheral vein by measuring changes in light absorption
Blood gas	Optical sensor	Measuring difference of absorption ratio of red and infrared light
Luminous intensity	Photodiode, phototransistor	To measure the intensity of ultraviolet rays, solar rays
Communication devices (like remote control)		
Color	CCD	Color and image sensing
Position	GPS sensor, RFID, ultrasound/infrared sensor	Absolute and relative positioning
Sound	Microphone	To record sound or measure noise level
Gas, aroma	Chemical sensor	Environment measurement
Humidity	Hygrometer	Environment measurement
Solar ray	Flexible solar cell	Energy source

the skin to reduce the electric resistance and to fix the electrode on the skin. However, coupling gel is difficult to use in clothing or in the daily living environment. Many researchers have been seeking alternatives that would not cause inconvenience such as skin trouble to patients, and the dry electrode and the capacitive-coupled electrode have been suggested (Ko et al. 1970; Burke and Gleeson 2000; Lee, Pearce, and Hib 2004; Karilainen, Hansen, and Müller 2005; Jang et al. 2007; Sung et al. 2007b). The dry electrode does not use coupling gel and measures bio-signals literally in dry state.

TABLE 4.2
Materials for Textile Electrode

Material	Merits	Demerits
Conductive rubber	High conductivity	Low flexibility
	Easy to shape and cheap	Poor air and liquid permeability
Silver-coated polymer foam	High conductivity	Poor washability
	Easy to shape and flexible	Poor air and liquid permeability
Metal-coated or sputtered fabrics	Fabric material	Poor washability
	High conductivity	Metal oxidation
Woven metal fabric	Easy to control conductivity (mixed blended spinning)	Difficult to handle Skin irritation
Woven conductive polymer fabric	Fabric material	Low conductivity

Conductive rubber, silver-coated polymer foam, metal-coated or sputtered fabrics, and woven metal fabric can be used as electrodes in smart clothing (see Table 4.2). Lately, fabrics using conductive polymers like polyaniline or polypyrrole have been developed and these fabrics are being used for electrodes or conductive materials.

Skin impedance can differ in accordance with skin conditions and electrical source, but it is generally known to range from 200 Ωcm^2 to 93 $K\Omega cm^2$ with a power of 60 Hz as a criterion (Enderle, Blanchard, and Bronzino 2005). Except for some special cases, the resistance and conductivity of the electrode should be lower than the skin impedance no matter what kind of material is used. Therefore, electric conductivity should be given the top priority in designing textile electrodes. Materials with low conductivity are weak to electrical noise, caused by impedance mismatching or electromagnetic interference, and make the acquisition of signals difficult. If the material is not flexible enough, it reduces the user's satisfactory feel of wearing and increases the noise caused by motion. Moreover, the textile electrode should do no harm to the human skin. Stainless steel yarn is liable to stimulate skin with its cut filaments, which causes stinging pain, whereas metals like nickel may cause skin irritation (Agency for Toxic Substances and Disease Registry [ATSDR] 2005). Therefore, one should be careful in electing the conductive materials. We need to pay attention to minute details not only in selecting conductive materials but also in making the clothes. Contact impedance of the dry electrode is much greater than the wet electrode, and the dry electrode is weak to various noises including the user's motions or radio waves. The easiest way to reduce contact impedance is to increase clothing pressure between the electrode and the body surface. For this purpose, it is common to use a chest strap structure for the ECG electrodes, but chest straps exert pressure on the chest and can cause discomfort and pain when worn for a long time. Now is the time to develop new concepts of clothes design that consider the user's convenience and the quality of signals.

Capacitive-coupled electrodes offer a new technique to measure ECG in non-contact state (Taelman et al. 2006; Dias et al. 2005). Strictly speaking, they are different

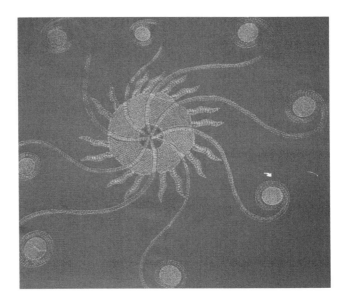

FIGURE 4.5   Embroidered electrode.

from textile electrodes, but are easy to apply to clothes, and the user's activity does not have to be restrained. Application of capacitive-coupled electrodes is highly recommended for smart clothing for such advantages.

In addition, the electrode can be made by embroidering conductive yarn on textile. Figure 4.5 shows an example of an embroidered electrode, which demonstrates that it is possible to make sensors by applying clothes manufacturing techniques alone. Embroidery can compose an electrode in beautiful patterns. It is also possible to make an electrode using Jacquard loom. Thus, the traditional textile techniques can also be effectively used for making sensors if they are applied in a happy combination with appropriate materials and methods.

Touch sensor is another field of electrode application whose operations principles are similar to capacitive-coupled electrode. It measures the changes of capacitance when the electrode is touched with a finger. The weakness of the touch sensor is its sensitiveness. It is so sensitive that it is also erroneously activated when it is touched by objects other than the finger. Except for this problem, the touch sensor is expected to be applied frequently to smart clothing since it is easy to make and quite suitable to the clothing environment.

Finally, the electrode is a sensor that receives signals, and at the same it can be used as an actuator for electrical stimulation. In the medical field, electrical stimulation is generally used for physical therapy. The electrode has also been used for functional electrical stimulation (FES) in the field of rehabilitation engineering. Researchers should be attentive to the following points in designing the electrode for electrical stimulation of the human body. Unlike the wet electrode that is being used in the medical field, the textile-based dry electrode is not evenly in contact with the human skin. Since electricity by nature flows to the path with the lowest resistance, the electric current tends to concentrate at the point where contact impedance is the lowest. When electric current flows into the resistors, heat is generated and electrical current beyond a certain value can inflict skin burns. Therefore, safety measures such as restricting the electrical current or developing a method of maintaining an even contact impedance must be taken in advance when the dry electrode is used for electrical stimulation.

4.3.3 TEXTILE PRESSURE SENSORS

Pressure is a physical quantity that is widely used not only in industries but also in our daily life. As can be seen in Table 4.3, pressure is itself an important physical quantity, but it can be processed into various forms of information to be used in daily life. A scale is a typical daily life apparatus in which pressure sensor is applied. When gravity acts on an object, pressure is added to the scale, and this pressure is converted to weight. Thus, the output of the pressure is sometimes used as the numerical value of pressure itself, and sometimes it is processed into various secondary physical quantities or information. When touch pressure is extracted by the pressure sensor, it can be converted to on/off signals of the switch.

Mechanical pressure sensors, strain gauges, and semiconductor piezoresistive or piezoelectric sensors are ordinary pressure sensors. Some kinds of film sensors can be suitable for the clothing environment, but other types of sensors are either too big or too hard to be applied directly to clothes. Therefore, we need to design special sensors that are suitable to the clothing environment. Various pressure sensors that have been researched to date include textile sensors that use textile structure (Sung et al. 2007a), the piezoresistive sensor (Peratech 2008), the capacitive pressure sensor (Meyer, Lukowicz, and Troster 2006; Sergio et al. 2002), and the flexible plastic optical fiber (POF) sensor (Rothmaier, Luong, and Clemens 2008; Tao 2002).

Coating or printing the pressure sensing material on the surface of fabric is the easiest way to measure the pressure applied to the fabric (Figure 4.6). For piezoresistive substances, conductive polymer compounds or elastomer that contains metallic particles can be used. This type of sensor has the advantage of measuring the relatively weak changes in pressure. On the other hand, use of a keypad is so sensitive to body movement or even to very slight friction that it is liable to erroneous operation.

TABLE 4.3
Application and Information Processing of the Pressure Sensor

Application	Information Processing
Weighing scales	Gravity $\rightarrow$ pressure $\rightarrow$ weight
Switch	Touch pressure $\rightarrow$ threshold $\rightarrow$ on/off
Respirometer	Respiration $\rightarrow$ chest volume $\rightarrow$ clothing pressure $\rightarrow$ respiratory rate
Gesture measurement	Acceleration by movement $\rightarrow$ dynamic gesture
	Changes of body surface $\rightarrow$ piezoresistive change $\rightarrow$ dynamic/static gesture
Accelerometer, vibroscope	Pressure change by inertia $\rightarrow$ acceleration/vibration

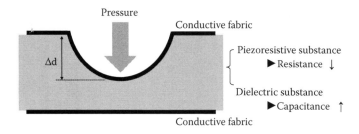

FIGURE 4.6 Principles of piezoresistive sensor and capacitive pressure sensor.

As in all sensors, we need the design of sensors that is accurate for the range of pressure we desire to measure. In other words, the sensor's exact operating range should be defined clearly before designing the sensors.

Another type of sensor is piezoelectric material made flexible by coating it on thin film. The piezoresistive sensors described in the previous paragraph measure changes in volume by converting them into electric resistance, and the piezoelectric sensor is a material that generates electricity. The piezoresistive sensor needs a process in which changes in resistance are converted to difference of potential, but the output of a piezoelectric sensor can be input directly into the analog circuit.

A capacitive pressure sensor uses the principle of a capacitor of electronics engineering (Figure 4.6). Capacitance is proportional to permittivity (dielectric constant) and the area of each metal plate, and inversely proportional to the distance between the plates. A capacitive pressure sensor applies this principle and makes a 3D structure by attaching conductive fabric on both the isolating layer and the supporting layer. When pressure is applied in orthogonal direction to the sensor, changes occur in the intervals of the supporting layer, which in turn changes the capacitance. A capacitive pressure sensor is sensitive enough to measure minute pressures, but it is weak to electric and mechanical noises of the external environment. Therefore, in designing the capacitive pressure sensor, proper material that fits the purpose and use of the sensor should be selected, and the size and shape should be defined clearly.

There is an example of textile pressure sensor made of flexible POF (Rothmaier, Luong, and Clemens 2008; Tao 2002). Changes in pressure cause changes in form and shape of plastic optical fiber, which in turn plays the role of gatekeeper for the flow of light. This effect is implemented in the textile pressure sensor. In general, POF does not change its form or shape in response to pressures applied in a radial direction. Therefore, we need a material that responds with physical changes to pressures exerted in a radial direction.

To measure horizontally exerted force, you might consider a textile sensor that uses the strain gage or the structure of fabric. Let us take the case of measuring human motions, for example. When the human leg bends, many changes occur in the clothes worn by that person. The biggest change occurs in the fabric in the vicinity of the knee. It stretches in a manner shown in Figure 4.7. The stretch means that the fabric structure has changed.

A motion sensor that uses the structure of fabric is a fabric that is capable of changing the mechanical variations of fabric into a measurable value, that is, an

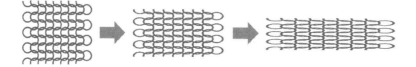

FIGURE 4.7 Transformation of textile structure by external forces.

electrically variable quantity. Knitted fabric has a loose structure when no external force is exerted, but it stretches and changes into a tight-knit dense structure. Conductive textile is needed to convert this into an electrical quantity. When conductive fabric is knit and it is loosely structured, most electric current flows through the thread paths. If the fabric is knit with one thread, the resistance at both ends of the thread is proportional to the length of the thread. If the knit fabric is stretched by external force, the space between threads is narrowed, and this leads to reduction of contact resistance or contact of one thread with another. When contact increases between threads, various current paths are generated. This causes the resistance to go down by composing a parallel resistance circuit. Thus, if resistance is measured at both ends of a fabric sensor, the value that changes with motion can be obtained. Although resistance value does not vary linearly in accordance with the angle of the leg, we can get the correlation curve for the angle and resistance through repeated experiments.

4.3.4 OTHER SENSORS

The textile-based sensor has been attracting much attention in the past 20 years. However, sensors that are made by the textile manufacturing process are not so many. Although sensors are being miniaturized thanks to advances in nano-technology, many problems remain to be solved before miniature sensors can be woven into fabrics. To satisfy this demand, it is necessary to combine the existing technologies. A good example is the recent attempt to mount miniaturized electronic chips on the flexible printed circuit board (FPCB) to be applied to the clothing environment. Compared to the flexibility of fabrics, FPCB is still very stiff but it could play the role of a good alternative by serving as a buffer between solid electronic chips with flexible fabrics. It is easy to compose an array structure like tactile sensors on the FPCB (Hwang and Kim 2005). Flexible solar cells are also a material that can be applied to clothing for the same reasons (Schubert and Werner 2006).

Another method that deserves research effort is to install miniaturized sensors on fabrics. Some thermocouples available in the market are thin and small enough to be woven directly into fabrics. If only the method of interconnecting the sensors and fabrics is developed, these sensors can be directly applied to the clothing environment. We may be able to find ways to apply even somewhat big sensors to clothing. The clothing environment uses a great deal of accessories such as buttons. There are many ways to put the buttons on the clothes. These methods should be explored for the possible ways to install sensors on the clothes. These methods may enable us to use almost all the electronic sensors in the clothing environment until a genuine textile sensor is developed.

4.4 TEXTILE-BASED ACTUATORS

In general, an actuator in engineering refers to a device that converts energy or electrical signal into mechanical motion, but in the realm of smart clothing, the actuator may better be defined as a component capable of delivering senses, information, or energy through a wearable system or by the user's intention. Sensor and actuator are transducers in that they change the form of energy. Some electronic parts actually possess the functions of both the sensor and the actuator at the same time. For instance, the photo diode can be used either as a light-detecting sensor or as an actuator that emits light like light-emitting diode (LED). Theoretically, speakers too can play the role of microphones. The sensor is an input part of the system while the actuator plays the role of an output device, which provides actual services. When a sensor perceives its surroundings, information is generated in the processing unit. However, actual services are provided in the actuator. An actuator in smart clothing can provide the following stimulations and services:

Stimulations
- Senses: sight, hearing, smell, touch
- Information: sign, warning, text, graphic
- Energy: heating, cooling, enhancing muscular functions, stimulating with electricity or light

Services
- Knowledge service: memory aid, context awareness, providing information and knowledge
- Communication service: exchanging ideas and opinions, social networking, media service
- Healthcare and safety service: health maintenance, treatment of diseases, aids for bodily functions, early warnings, diagnosis, detection of diseases
- Emotional service: offering pleasant and comfortable environment according to emotional state, sensibility communication

A stimulation can provide multiple services and the same stimulation can offer different services to different people. The actuators can be divided into passive actuator and active actuator. The passive actuator operates by absorbing energy from environmental changes or stimulations while energy is supplied to the active actuator from the external source and it can be controlled by the user's intention.

The passive actuator is a material that reacts to stimulations by using the material's own peculiar property, although the power source is not provided. Since the material's peculiar property plays the role of a program, the passive actuator carries out predetermined programs following the changes in the circumstances. Consequently, it operates by itself without interventions from humans or systems. Its installation on clothing does not require complicated procedures.

Most active actuators are controlled by an electronic system, and the power source is electricity. Although precision control is possible with its multiple functions, the active actuator has many problems to solve before it can be installed on clothing.

Battery for the supply of electrical power is also a factor that interferes with the design of the clothes. The passive actuator and the active actuator both have advantages as well as weaknesses. There are times when both are used simultaneously for a single function.

Some examples of textile-based actuators or potential actuators applicable in the clothing environment are described below.

4.4.1 SHAPE MEMORY MATERIALS

Shape memory material may be considered for a material to induce mechanical variations in clothing. There are two types of shape memory material: one changes its shapes with temperature changes, and the other changes its shapes with electrical stimulations. Shape memory alloy (SMA) and shape memory polymers (SMP) are materials that change their shapes as temperature changes. SMA and SMP can carry out either passive control or active control depending on the purpose of its use. The material that performs a passive action at a certain temperature needs to have a proper temperature characteristic. This kind of material can be used, for example, in designing clothes that widen their openings for ventilation when the temperature within the clothes rises.

Electroactive polymer (EAP) is another material that is drawing attention recently. Because shape changes occur at a relatively low voltage, electroactive polymer is often being used as a material for tactile display or artificial muscles (Bar-Cohen et al. 2007; Sahoo, Pavoor, and Vancheeswaran 2001). EAP is expected to be more efficient than SMA or SMP because it is electrically controllable. Bi-metal board with two different kinds of metal is also used to transform shape as temperature changes.

Active control is used when a particular purpose calls for specific action. In the case of SMA, electric current can flow through it because of the nature of metallic material, and heat is generated when electric current flows through metals. Thus, it is possible to control variations of shape by controlling the current that runs through SMA. In addition, there have been other research efforts (Lendlein et al. 2005; Mohr et al. 2006) on active control of shapes of SMA and SMP by applying electromagnetic fields or light.

4.4.2 THERMOREGULATION MATERIALS

Temperature control is one of the important functions of clothes. Thermoregulation by smart clothing is being approached from two angles. One is the research on textile material that can radiate heat beyond its heat preservation function, and the other is seeking ways to enhance the comfort of the wearer by controlling the temperature of the clothes.

Most of the heating elements use the principle of Joule's heat, which is generated when electric current is passed through conductive material. Therefore, heating material is the most important factor in designing the heat-radiating textile. All conductive materials are heating elements in principle; however, the voltage supplied, the temperature of heat radiated, and the capacity of the battery are the problems to examine before selecting the heating material.

Heating elements are divided into two categories by shape: sheet-type heating element and wire-type heating element. A sheet-type heating element is usually made by adsorbing conductive material to the fabric. A wire-type heating element can be made with conductive thread or metal yarn. The next consideration is whether or not the material is suitable for clothing. The heating element to be applied to clothing should be soft and flexible and be able to endure washing. Thus, material like nichrome wire, which is generally used in the electrical industry, is not suitable for clothing. Last but not the least important is the safety factor. How to connect the heating element and the power source is no less important than how to make the heating element energy efficient. In other words, the electro-textile interface is the most important issue to solve. If the connection between heating element and power line becomes loose, the wearer can suffer from fire or burns caused by abnormal heating. Therefore, a completely safe method should be secured that guarantees flawless electrical connection.

Phase change material (PCM) is most frequently used for enhancing comfort through temperature control (Mondal 2008). If a textile material can absorb or emit heat at a certain temperature, it would be useful in making comfortable clothing. When temperature goes up, the PCM melts and absorbs heat, and when temperature goes down, PCM solidifies and emits heat. The controllable temperature range is determined by the PCM's crystallization temperature. The procedure of PCM application to textile is as follows. First, PCM such as paraffin or hydrated inorganic salt is inserted into a tiny microcapsule. Then, the PCM microcapsule is attached to the fiber while the textile is produced. The capacity of heat absorption or emission is determined by the thermodynamic features and amount of the PCM. It is important to select the best heat absorbing material at normal temperature within the clothes (15–35°C). The textile's heat capacity would increase if a greater amount of PCM is used. However, there is a limit to the size of the microcapsules. Still another problem to solve is how to make small and sturdy capsules and how to attach them firmly to fibers.

4.4.3 LUMINESCENT AND COLOR-CHANGING MATERIALS

Light is the most familiar and yet the fastest reacting stimulation to humans. The textile-based luminescent and color-changing material is applied in the following fields.

- Safety: fatigue clothes, special uniforms for policemen and fire fighters, road signs, security products
- Sports: jogging suits, clothing for cyclists, climber's clothes, etc.
- Entertainment: children's clothes, clothes for juniors, party dresses, etc.
- Sensibility and medicine: emotional clothing, sterilized clothes, light therapy, etc.
- Interior decoration: curtain, wallpaper, table cloth, etc.

Luminescent and color-changing materials (Project TeTRInno SmarTex 2007) are used for a variety of stimulating effects as shown in Table 4.4.

If these effects are reversed, they can be used as sensors. For instance, materials using thermochromism functions as an actuator, which changes colors in response to

TABLE 4.4
Luminescent and Color-Changing Materials

Category	Phenomenon	Stimulus
Luminescent material	Photoluminescence	Light
	Opticoluminescence	Conduction of light
	Electroluminescence	Electricity
	Chemiluminescence	Chemical reaction
	Triboluminescence	Friction
	Sonoluminescence	Sound
	Radioluminescence	Ionizing radiation
	Crystalloluminescence	Crystallization
Color-changing materials	Photochromism	Light
	Thermochromism	Heat
	Electrochromism	Electricity
	Piezochromism	Pressure
	Solvatochromism	Liquid or gas
	Halochromism	pH
	Tribochromism	Friction

the environmental temperature, and at the same time it has the sensor's function of sensing the body temperature in response to color changes.

Photoluminescence material, opticoluminescence material, and electroluminescence material are three major luminescent materials that can readily be applied in the manufacturing process of textile or clothing. Photoluminescence material mostly uses the dyeing process. Because it just absorbs and re-radiates photons, it reacts passively by the light environment rather than reacting actively. Photoluminescence material can be used for safety in dark places or in party dresses for entertainment. POF is mainly used for opticoluminescence material. POF delivers light signals effectively, and for that reason POF is used a great deal in fiber-optic communications. The optical fiber core is clad to prevent the light leaking out of the optical cable. However, for visual effects in clothing, the light should leak out of the optical fiber. Therefore, the cladding layer needs to be damaged on purpose. Two kinds of processing may be considered: the cladding layer can be either physically damaged or chemically corroded. It is possible to draw patterns or logos desired through partial damaging.

For the light source, LED or low-power laser diode can be used. The advantage of this method lies in the fact that the system can display a great variety of colorful expressions by controlling colors and light intensity. The disadvantage is the difficulty in obtaining a sufficient amount of light because light is transmitted indirectly through POF.

Still other problems remain to be solved. POF is not yet flexible enough to be applied to clothing, and it is still not easy to connect many POFs to the light source. LED or flexible electroluminescence (EL) sheet may well be considered for electroluminescence material for smart clothing.

The LED is a light-radiating component with various sizes and shapes, and it is readily available. Since it has high energy-converting efficiency, the LED can be activated with a small amount of electric current, and one LED can express almost all colors. LED is either attached to textile directly or mounted on FPCB. Thus, research is called for in search of effective ways to attach light-emitting diodes on clothing for various colorful expressions.

Recently, flexible EL sheet (Elastolite n.d.) has been developed, and it is being applied to clothing. It can display a variety of elaborate colors, which can be used not only for safety but also for esthetic expressions. Though it is not as bright as LED, the flexible EL sheet is much brighter than the light-emitting textile that uses POF. Because the flexible EL sheet uses elastic material such as polyurethane, it is very soft. Moreover, since it is made washable, the flexible EL sheet is widely applicable in many fields. There is no problem in attaching it to clothing because the common bonding technique is used, which is widely used in clothes manufacturing. The only thing that calls for special attention is that it needs a device called an inverter if it is to be activated by battery, and therefore it should be waterproof.

Color-changing materials, known as chameleon fiber, change colors as the external environment changes. Thermochromic material is the technique most widely applied in clothing. Since it changes colors as the temperature changes in the environment, thermochromic material is used in manufacturing casual suits for young people or outdoor clothing. Thermochromic material can be used in combination with heating fabric to induce active color changing according to the user's wishes.

4.5 CONCLUSION

Research on smart and intelligent textile began at the end of the 20th century, and it has made a quantum leap in the 21st century. Research interest has been shifting from materials with passive and active functions to the development of materials with intelligent functions. Originally started by researchers in textile engineering, now researchers from various fields are participating in intelligent textile research including clothing designers, engineers from all the engineering fields, medical fields, psychology, and liberal arts. If the 20th century was the age of specialization and compartmentalization, the 21st century is headed toward the age of fusion and integration of diverse fields of learning. Scholars have realized only recently that many technologies are hidden behind the barriers between disciplines that can be integrated for mutual benefits. Consumer demand is the driving force behind this flow of positive trends. Sloughing off the age of mass production and mass consumption, a new age has come—an age of made-to-order products and services that respect personal individuality. Furthermore, the infrastructure of information and communication is changing the daily life-style completely. The arrival of the ubiquitous age means that the environment is information and services at the same time. In this kind of environment, the ability to sense and enjoy the information and services that are spread out there waiting to be utilized is the absolute requirement for future life.

The ubiquitous age requires the sixth sense in addition to the traditional five human senses. The sixth sense could be all kinds of abilities to communicate with the ubiquitous environment. The most urgent task before us is to find out what the sixth sense is and design clothes that apply this sixth sense. This is very creative work that transcends the conventional way of thinking. Questions like, "What kinds of materials are needed and what technologies should we apply?" can wait until the question on the sixth sense is answered. There are a countless number of sensors and actuators implemented by the existing technologies. By far, there are more sensors and actuators that cannot be applied to clothing now than can be applied, but if there is demand in the market, it is possible to convert many technologies to those that can be applied to clothing.

This chapter reviewed the sensors and the actuators that can be applied to smart clothing. A great variety of sensors exist, and the sensors for the same purpose of use have applied different materials as well as different technologies. Therefore, this chapter described only several examples that are typically representative. To emphasize once again, it is possible to use any material or element as a sensor or actuator if it receives energy and transduces itself into other form or shape. This means that the physical, chemical, and electrical characteristics of all fibers, fabrics, and accessories for clothing should be reinvestigated. If it is not possible to measure those characteristics directly, an indirect way should be sought. The technology for the interface between electronic element and fabric could be found among the traditional sewing processes. If direct interface is difficult, then an indirect method or a rerouting method should be researched, and the search for an indirect method requires a greater variety of knowledge and complicated processes. This hidden technology is extremely difficult to find from the viewpoint of the researcher's own major field alone. The only way to make smart clothing intelligent is to boldly explore various fields of learning with a positive attitude and share ideas and thoughts with other researchers.

REFERENCES

Agency for Toxic Substances and Disease Registry (ATSDR). 2005. Public health statement: Nickel (CAS#: 7440-02-0). http://www.atsdr.cdc.gov/toxprofiles/tp15-c1-b.pdf (accessed September 1, 2008).

Bar-Cohen, Y., K. J. Kim, H. R. Choi, and J. D. W. Madden. 2007. Electroactive polymer materials. *Smart Materials and Structures* 16.

Baurley, S. 2004. Interactive and experiential design in smart textile products and applications. *Personal and Ubiquitous Computing* 8:274–81.

Burke, M. J., and D. T. Gleeson. 2000. A micropower dry-electrode ECG preamplifier. *IEEE Transactions on Biomedical Engineering* 47:2.

Dias, T., R. Wijesiriwardana, K. Mitcham, and W. Hurley. 2005. Capacitive fibre meshed transducers for touch and proximity sensing applications *IEEE Sensor Journal* 5:3.

Elastolite. n.d. http://www.oryontech.com/aboutEL.asp (accessed September 1, 2008).

Enderle, J., S. Blanchard, and J. Bronzino. 2005. *Introduction to Biomedical Engineering,* 2nd ed. Academic Press.

Food and Drug Administration (FDA). 2003. Cell phone facts: Consumer information on wireless phones. http://www.fda.gov/cellphones (accessed September 1, 2008).

Hwang, E. S., and Y. J. Kim. 2005. Flexible tactile sensor based on polyimide for shear force detection. Paper presented at 1st International Conference on Manufacturing, Machine Design and Tribology (ICMDT 2005), Seoul, Korea.

Jang, S. E., J. Y. Cho, K. S. Jeong, and G. S. Cho. 2007. Exploring possibilities of ECG electrodes for bio-monitoring smartwear with Cu sputtered fabrics. Paper presented at 12th International Conference on Human-Computer Interaction International (HCII) 2007, Beijing, China.

Karilainen, A., S. Hansen, and J. Müller. 2005. Dry and capacitive electrodes for long-term ECG-monitoring. Paper presented at the 8th Annual Workshop on Semiconductor Advances.

Ko, W. H., M. R. Neuman, R. N. Wolfson, and E. T. Yon. 1970. Insulated active electrodes. *IEEE Transactions on Industrial Electronics and Control Instrumentation* 17:195–98.

Lee, J. M., F. Pearce, and A. D. Hib. 2004. Evaluation of a capacitively-coupled, non-contact (through clothing) electrode or ECG monitoring and life signs detection for the objective force warfighter. Paper presented at the RTO HFM Symposium on Combat Casualty Care in Ground Based Tactical Situations: Trauma Technology and Emergency Medical Procedures, St. Pete Beach, USA.

Lendlein, A., H. Jiang, O. Jünger, and R. Langer. 2005. Light-induced shape-memory polymers. *Nature* 434 (7035): 879–82.

Linz, T. 2007. Enabling micro system technologies for electronics in textiles. Paper presented at Concertation WS on EC-Funded Projects on Smart Fabrics and Interactive Textiles (SFIT) and Consultation on Future R&D Challenges and Opportunities, Brussels.

McCraty. R., M. Atkinson, W. Tiller, G. Rein, and A. D. Watkins. 1995. The effects of emotions on short-term power spectrum analysis of heart rate variability. *The American Journal of Cardiology* 76 (14): 1089–93.

McLuhan, M. 1968. *War and peace in the global village*. Bantam

Meyer, J., P. Lukowicz, and G. Troster. 2006. Textile pressure sensor for muscle activity and motion detection. Paper presented at the 10th IEEE International Symposium on Wearable Computers, Montreux, Switzerland.

Mohr, R., K. Kratz, T. Weigel, M. Lucka-Gabor, M. Moneke, and A. Lendlein. 2006. Initiation of shape-memory effect by inductive heating of magnetic nanoparticles in thermoplastic polymers. *Proceedings of the National Academy of Sciences of the United States of America* 103 (10): 3540–45.

Mondal, S. 2008. Phase change materials for smart textiles—An overview. *Applied Thermal Engineering* 28:1536–50.

Peratech. 2008. The science of QTC. http://www.peratech.com/science.htm (accessed September 1, 2008).

Project TeTRInno SmarTex. 2007. State of the art in smart textiles and interactive fabrics. http://www.mateo.ntc.zcu.cz/doc/State.doc (accessed September 1, 2008).

Rothmaier, M., M. P. Luong, and F. Clemens. 2008. Textile pressure sensor made of flexible plastic optical fibers. *Sensors* 8:4318–29.

Sahoo, H., T. Pavoor, and S. Vancheeswaran. 2001. Actuators based on electroactive polymers. *Current Science* 81 (7).

Schubert, M. B., and J. H. Werner. 2006. Flexible solar cells for clothing. *Materials Today* 9 (6).

Sergio, M., N. Manaresi, M. Tartagni, R. Guerrieri, and R. Canegallo. 2002. A textile based capacitive pressure sensor. Paper presented at Sensors 2002, Orlando, Florida, USA.

Sung, M. S., K. R. Baik, J. Y. Cho, K. S. Jeong, and G. S. Cho. 2007a. Characteristics of low-cost textile-based motion sensor for monitoring joint flexion. Paper presented at the 11th IEEE International Symposium on Wearable Computers, Proceedings of the Doctoral Colloquium, Boston, Massachusetts, USA.

Sung, M. S., Y. J. Yang, J. Y. Cho, K. S. Jeong, and G. S. Cho. 2007b. Comparing signals of textile-based ECG electrodes with signal of AgCl electrode. Paper presented at International Conference on Intelligent Textiles (ICIT) 2007, Seoul, Korea.

Taelman, J., T. Adriaensen, A. Spaepen, G. R. Langereis, L. Gourmelon, and S. Van Huffel. 2006. Contactless EMG sensors for continuous monitoring of muscle activity to prevent musculoskeletal disorders. Paper presented at the first Annual Symposium of the IEEE/EMBS Benelux Chapter, Brussels, Belgium.

Tao, X. 2002. Sensors in garments. *Textile Asia* January: 38–41.

Van Langenhove, L., and C. Hertleer. 2004. Smart clothing: A new life. *International Journal of Clothing Science and Technology* 16 (1/2): 63–72.

Weber, W., R. Glaser, S. Jung, C. Lauterbach, G. Stromberg, and T. Sturm. 2003. Electronics in textiles: The next stage in man machine interaction. Paper presented at the 2nd CREST Workshop on Advanced Computing and Communicating Techniques for Wearable Information Playing, Nara, Japan.

Weiser, M. 1991. The computer for the 21st century. *Scientific American* 265:66–75.

Zhang, X., and X. Tao. 2001. Smart textiles: Passive smart. *Textile Asia* June: 45–49. Smart textiles: Active smart. *Textile Asia* July: 49–52. Smart textiles: Very smart. *Textile Asia* August: 35–37.

5 Integration of Plastic Optical Fiber into Textile Structures

Moo Sung Lee, Eun Ju Park, and Min-Sun Kim

CONTENTS

5.1 INTRODUCTION

One of the noticeable trends in textile industries is to integrate electronic wires or optical fibers (OFs), which can transmit electric or optical signals, into the textile structure. Such textiles can be used as wearable electronic or optical circuits if electronic or optical devices are coupled to them. They also provide new opportunities based on the development of new technologies to successfully miniaturize and embed electronics, optics, and sensors into fabrics and garments.

Most mobile devices that can be connected to wearable interactive textiles are based on electronics. Textiles with electronic wires have been developed from the beginning of research in this field because they are sufficiently tough enough to withstand the textile manufacturing process. However, electrical wires cannot avoid the problems caused by electromagnetic interference, which can lead to malfunction of electronic devices. For example, electromagnetic fields can be critical for people monitoring vital signals. Furthermore, many people are reluctant to have wire against their body.

OFs are able to carry light along their length and are immune to electromagnetic fields. Specifically, fiber made of plastic or polymer, called as plastic (polymer) OF (POF), is tougher than glass OF (GOF) and can be integrated into the fabric using conventionally standard textile machines. This was achieved by Dr. S. Jayaraman, who invented a way of making a "wearable motherboard" or smart shirt. This innovation opened the possibility of developing a flexible, wearable, and comfortable garment with sensors for monitoring a variety of vital signs, including heart rate, electrocardiogram, body temperature, and pulse oximetry (Park, Mackenzie, and Jayaraman 2002). His invention has sparked a worldwide boom in the developmental research of fabrics integrated with POF (Harlin, Mäkinen, and Vuorivirta 2002; Im et al. 2007; Clemens et al. 2003). OF also has other advantages such as the ability to easily and inexpensively connect to other optical devices such as light source, connector, and photo detectors. Moreover, since POF is much softer than conductive metals, its use in interactive textiles renders them more human friendly than electrical wire.

Even though integrating POF into the textile structure is a novel approach, many problems remain for its commercialization. As pointed out by Graham-Rowe (2006), one of the biggest challenges is the actual weaving, stitching, and knitting of the fabrics because POF is easily broken, which has limited its reliability and durability. Moreover, only a few applicable optical devices for connection have been developed.

In this chapter, we present the basic properties of POF, the manufacture of side-emitting POF, the integration of POF into the textile structure, and then a brief review on the application of the fabric being integrated with POF, focusing on the applicability and limitation of POF as a fabric component. The terminology for describing fabric woven with POF is not clearly defined at this moment. "Photonic textiles" is used to refer to fabrics that contain lighting systems and can serve as displays or other uses, even though they include the fabrics that are integrated with flexible arrays of multicolored light-emitting diodes (LEDs) (Philips 2008). In this study, "photonic textiles" is used when POFs woven into the fabrics act as a light-illuminating medium.

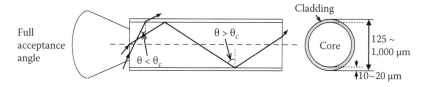

FIGURE 5.1 Light guiding in a typical PMMA-core POF.

5.2 BASICS OF POFS

5.2.1 POF AS A SIGNAL TRANSMITTING MEDIUM

5.2.1.1 Structure and Materials

Like most OFs, POF consists of two parts: the core and the cladding, as shown in Figure 5.1. Even though light does not propagate through the cladding, it reduces scattering loss at the surface of the core, protects the fiber from surface contamination, and adds mechanical strength. To guarantee that light is kept in the core, the total internal reflection condition should be met:

$$n_{core} > n_{cladding}$$

where n_{core} and $n_{cladding}$ are the refractive indices of the core and cladding, respectively.

The critical angle, θ_c of OFs determines whether an internal ray with the incidence angle θ will be reflected back into the core or refracted from the core into the cladding. Only when $\theta > \theta_c$ are the rays completely reflected. The critical angle is determined by the difference in the index of refraction between the core and cladding materials:

$$\theta = \arcsin\left(\frac{n_{cladding}}{n_{core}}\right)$$

The critical angle determines the acceptance angle of the fiber, often reported as a numerical aperture (NA):

$$NA = \sqrt{n_{core}^2 - n_{cladding}^2}$$

and full acceptance angle $= 2 \times$ acrsinNA. A high NA allows light to propagate down the fiber in rays both close to the axis and at various angles, allowing efficient coupling of light into the fiber. If the fiber is bent, the propagation path of light and thus the incidence angle is changed and thus light with $\theta < \theta_c$, will refract as it hits the sides of the fiber.

POF is made completely of polymer. The materials used for the core are almost all poly(methyl methacrylate) (PMMA, n = 1.49), while fluoropolymers are used as cladding materials for the PMMA core. The details on the chemical structure of the cladding materials are not well known, but the refractive index decreases with increasing fluorine content (Lekishvili et al. 2002). The refractive index of fluoropolymers depends on the fluorine atom content and is in the range of $1.37 < n_{cladding} < 1.44$, which corresponds to an NA of POF in the range of $0.38 < NA < 0.59$. Polystyrene (PS, n = 1.59) may be used as the core with a cladding of PMMA or styrene-methyl methacrylate copolymers. Although PS-core POF shows relatively low attenuation compared to PMMA, it is very brittle and only used for decorative purpose.

Since bare POF is susceptible to environmental damage, it is usually covered with jacketing materials such as polyethylene, poly(vinyl chloride), and polyamide. The outer diameter of the jacketed POF, termed the POF code, is usually 2.2 mm and much larger, in the order of tens to hundreds of micrometers, than that of conventional yarn being used for textile manufacturing.

5.2.1.2 Attenuation and Bandwidth

Attenuation is the gradual loss in intensity of light through the fiber and thus controls the distance an optical signal can travel. Attenuation is measured in decibels (dB), which are logarithmic units commonly used to relate input power to output power, or dB/km^{-1}. Attenuation of POF is caused by several factors, which can be classified into the inherent loss in the material itself and the external loss from the manufacturing process. In photonic textiles, additional losses from the damage suffered by POF during manufacturing or wearing and macro- or micro-bends of POF in textile structure should be considered. Attenuation of PMMA increases considerably with the wavelength and is very high in the near infrared due to the harmonic generation of C–H stretching, providing acceptable attenuation only in the visible spectrum from 350 nm up to 750 nm. Commercial PMMA fibers have an optical loss in the order of 140 dB/km in red and 80dB/km in green. Although it is much higher than that of GOF, which is below 0.5 dB/km near the infrared range, POF is sufficiently useable for short links such as wearable photonic textiles, where the length of the fiber does not exceed several tens of meters.

The bandwidth of OF is defined as its information carrying capacity and is closely related to the dispersion, i.e., the spreading of the optical pulse as it travels along the fiber. Since the bandwidth of a POF link depends on the rise time of the light source and detectors, as well as POF itself, bandwidth × length is usually used as the measure of a fiber's ability to transmit high-speed signals. In the case of large-diameter POF, a major limitation on the bandwidth is modal dispersion, in which different optical modes propagate at different velocities and the dispersion grows linearly with length. Such intermodal dispersion can be reduced by lowering the NA of POF or designing a new fiber with a graded index profile so that the refractive index is high in the center but reduces outwards. The bandwidth of a typical PMMA fiber is limited to 40 MHz × 100 m, but that of graded-index POF is over 2 GHz × 100 m. The details on the data transmission of various OFs are given in a reference (Ziemann et al. 2005). Up to now, the combination of step-index POF and 650 nm LED is expected to be the lowest cost solution for the POF optical link, despite its low bandwidth. Since high bandwidth fiber is suitable only for applications such as gigabit ethernet and digital video interface, this solution might be sufficient for phonic textile applications.

5.2.1.3 Mechanical/Physical Properties

POF should not be broken during the textile manufacturing processes, such as weaving or knitting, or while wearing it. The change in the optical properties should also be minimized after being integrated into the textile structure. Therefore, the fiber fabricated into textile structure is in a very different environment from that when it is used in the form of a code or cable for data communication. The damage suffered by the fiber causes extra optical losses and unfortunately weakens the POF to the extent that it is finally broken. In order to integrate the POF into the textile structure successfully, its mechanical properties such as tensile strength, abrasion by friction, bending, and fatigue need to be considered. Unfortunately, however, most data provided from manufacturers are for POF code, not for bare POF (Marcoe 1997).

Figure 5.2 shows the tensile properties for a polyester yarn (150D/48F) being used to construct photonic textile and for a bare POF with a diameter of 0.3 mm. The POF diameter is almost twice as large as that of the polyester yarn. The data for the mechanically damaged POF for side-illumination are also given for comparison. The PMMA fiber shows typical ductile fracture behavior and has a tensile strength comparable to that of the polyester yarn, indicating that bare POF can be woven or knitted using a standard machine without special attention. A tension of less than 10% of the permanent deformation point is generally recommended for normal use. However, the POF with surface flaws due to mechanical notching has slightly lower strength than the bare one has and also shows brittle fracture, which decreases the strength and elongation because the defects act as stress concentrators and thus accelerate the crack propagation. Since the thickness of the cladding layer is only

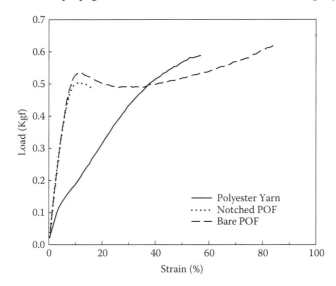

FIGURE 5.2 Representative tensile properties of PET yarn, bare POF, and POF mechanically notched for side-illumination. The tensile speed was fixed at 100 mm/min.

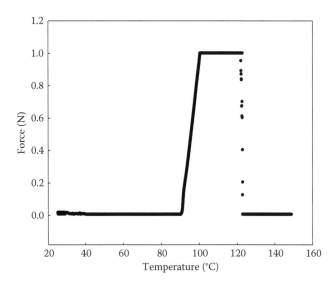

FIGURE 5.3   TMA shrinkage thermogram of a typical PMMA, 1.0 mm diameter POF.

around 10 μm, surface damage due to abrasion and other factors cannot be avoided during textile processing. This damage makes it difficult to mass produce photonic textiles without severely decreased optical and mechanical properties.

Figure 5.3 shows the thermal shrinkage for a 1-mm-diameter PMMA fiber, measured using a thermomechanical analyzer (TMA). Above 90°C, just below the glass transition temperature of PMMA, the fiber undergoes a rapid and irreversible shrinkage process due to the relaxation of the oriented polymer chains. Since thermal shrinkage determines the end-use potential of a given fiber, such as the safe ironing temperature, the maximum use temperature of POF of about 70°C must be considered. When POF is exposed above this temperature for a prolonged period, the dimensional stability and optical properties are significantly decreased.

Besides the tensile and thermal properties mentioned above, numerous mechanical properties such as repeated bending, flexing, torsion, impact strength, and vibration influence the reliability of POF and thus the resultant photonic textiles. Experimental results on these can be found on the websites of references (Asahi-Kasei 2008; Mitsubishi Rayon 2003; Toray 2008).

Table 5.1 lists the manufacturers of POF and its basic properties as provided by them. The major suppliers are all in Japan. More detailed information can be found from their websites given in the references section. Although not listed in this table, suppliers such as Nuvitech (Korea), Optimedia (Korea), Nexans (France), and Chromis Fiberoptic (USA) are all minor suppliers.

5.2.2   POF as Illuminating Medium

POF has larger core area fraction of over 90% and NA of around 0.5, allowing it to carry more light than GOF to the point of use. Applications such as image guide

TABLE 5.1
Major POF Suppliers

Suppliers	Trade Name	Characteristics	Applications
Asahi KASEI	Luminous™	φ_{fiber}: 0.5~3.0 mm NA: 0.5~0.6 Multi-core fiber Image fiber	Decoration Image transmission Light guide Data transmission
Mitsubishi Rayon	ESKA™	φ_{fiber}: 0.25~3.0 mm NA: 0.3~0.5 Bandwidth: 40~500 MHz/50 m	Illuminations Light guides Automotive Home network

Toray	Raytela™	φ_{fiber}: 0.25~3.0 mm
		NA: 0.32~0.63
		Bandwidth: ≥10 MHz/
		  km
			Light guide
			Automotive
			Broadband
			  communication

Asahi Glass	Lucina® (Asahi
	  Glass 2008)
		Graded index-type
		  perfluorinated
		  optical fiber
		Core/fiber diameter:
		  120/500 μm
		Bandwidth: >2.5
		  Gbps over 100 m
		Loss: 20~40 dB/km
		  (@ 850~1300 nm)
			High-capacity
			  transmission

Note: φ_{fiber}, fiber diameter; NA, numerical aperture.

and outdoor lighting take advantage of its light-carrying capability. Flexibility and ease of connection to light source are other advantages of POF lighting. Unlike the end-lighting POF, which transports light to one end of POF, side-illuminating POF distributes the light from the light source along the entire length of the fiber, as in neon lighting. The fabrics being integrated with such POF have various applications in fabric display, 2D lighting, and health monitoring sensors.

Side-emitting POF can be made in several ways by adding specific scatters or fluorescent additives into POF (Daum et al. 2002), by mechanically damaging the core-cladding interface, or by bending the fiber axis. The latter two approaches have potential for commercialization and are discussed below.

5.2.2.1 Mechanical Damaging of the Core-Cladding Interface

Structural irregularity or imperfection at the core-cladding interface or in the core changes the propagation path of light and allows light leakage. Such imperfection can be made on the surface of POF by mechanical notching (Koncar 2005), abrasion (Im et al. 2007), or sandblasting (Endruweit, Long, and Johnson 2008). Abrasion may also accidentally occur during weaving. Except for sandblasting, irregular emission pattern is obtained along the fiber length. As shown in Figure 5.2, however, the physical damage weakens the POFs and renders them difficult to weave or knit in the form of fabrics. One solution to this problem is to apply mechanical damage to the surface of the POFs after integrating them into the fabric. In this case, it is important to design the fabric with a structure that exposes the POFs to the surface of the fabric in order to maximize the light-emitting efficiency.

5.2.2.2 Fiber Bending of the Fiber Axis

OFs including POF are sensitive to bending. The bending sensitivity depends on the fiber diameter and NA. Optical losses increase with decreasing bending radii, and fibers with larger NA show higher losses at the same bending radii. Depending on the POF diameter, yarn density, and weaving pattern, multiple bends are introduced in the OF as it is woven into a fabric. When the incidence angle behind the bend exceeds the critical angle, the light ray escapes from the core and thus is emitted uniformly along the length. Since light escapes every time $\theta > \theta_c$, a part of the light rays might be emitted from the rear surface of the fabric, thereby decreasing the illuminating efficiency. The influence of the fabric structure and POF diameter on the bending losses and in photonic textiles has been discussed in some references (Harlin, Mäkinen, and Vuorivirta 2002; Atsuji et al. 2006; Endruweit, Long, and Johnson 2008). According to the Monte Carlo simulation results on the light transmission in OF woven fabric (Endruweit, Long, and Johnson 2008), the most significant transmission loss takes place in the initial region to be bent and bending losses increase when the bending radii are reduced. The side-emission intensity strongly depends on how the POFs are coupled to the light source. The most promising solutions are to use multiple light sources at each fiber end or to attach a reflector to one end (Spigulis 2005).

5.2.3 KEY COMPONENTS OF THE POF OPTICAL LINKS

Unlike laser diodes (LDs) and LEDs, POF itself does not emit light except for luminescent OFs and thus should be used together with other optical components. Figure 5.4 presents a schematic diagram of a basic optical link.

The transmitter contains a light source such as LED, LD, or vertical cavity surface-emitting laser diode (VCSEL) that converts an electrical current into an optical signal. Various types of light source, especially designed for POF, have been developed and are easily obtained from several suppliers. The high power of these light sources can damage the human eye or skin, especially for the case of LD, so care must be taken. All laser products are classified according to their

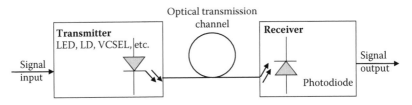

FIGURE 5.4   Optical components of a basic plastic optical link.

safety when they are exposed to the human eye. For example, class 2 with a power output below 1 mW, is safe for accidental viewing under all operating conditions (Lasermet Ltd. 2008). When photonic textiles are used for medical applications, the classification of the laser, as well as the wavelength and type of light source used, should be considered.

The receiver contains a photodiode or a PIN photodiode that converts the light back into an electrical signal, an amplifier that makes the signal easier to detect, and a discriminator that is able to recognize if the bit received is a low or a high. The fiber-optic cable carries the optical signal between them. When the POF is used as a light illuminating medium, the receiver is unnecessary. Optical components such as connector and splitter can be included in the optical transmission channel. The efficiency of an OF as a signal transmission channel is affected by the precision in transferring the light between the OF and other optoelectronic devices such as a photo detector or laser. The alignment tolerances are very small, thus requiring housings that can maintain precision alignment in service.

A number of connector systems have been developed and are available depending on the requirements for mechanical stability (Daum et al. 2002). However, most of them are for POF code with a jacket, not for bare POF and POF bundles, and thus they must be designed and produced according to their own purpose. The mechanical factors such as tension and vibration applied to them during wearing the photonic textiles should be considered for their design. Unlike GOF, POF connectors and connectorization are easy to install quickly with low-cost tools. By simple grinding/polishing or using a hot plate, a smooth end face of POF is obtained that can be cut with a razor blade.

5.3 INTEGRATION OF POF INTO TEXTILE STRUCTURES

5.3.1 WEAVING AND KNITTING OF POF

The majority of textile materials can be regarded as flexible flat sheets. The exceptions are lines, cord, ropes, and braids, which are one dimensional, and a few fibrous structures built up in three dimensions. Strictly speaking, even the flat sheets have an important three-dimensional structure within their thickness, and depart in a major way from two dimensions when they are allowed to buckle, bend, or drape.

As mentioned in the previous section, textile processing such as weaving can affect the mechanical and optical properties of POF. For the appropriate integration of POF, therefore, the processing parameters and final shape conditions of the textile structures need to be carefully identified.

5.3.1.1 Woven Structure

Woven fabrics are made by interlacing two sets of yarns at right angles to each other. The lengthwise yarns are known as warp yarns or ends, while the widthwise yarns are known as weft yarns or picks. Warp yarn in a fabric can be distinguished from the filling yarns in various ways. For example, warp yarns are usually thinner, possess more twist, are more compact, and have less stretch than weft yarns. The most important point in integrating POF into textile is that POF should be compatible with the warp or weft yarns, which are composed of textile. POF must be introduced into the structure just as the yarns are woven. In weaving machines, the yarns in the warp direction are pulled out from a roller that contains many individual strands of yarns. These yarns are then formed into a woven structure by passing a strand of yarn in the transverse or weft direction via a shuttle, which interlaces the yarns. The yarns in the weft direction are not continuous and their length cannot be very long since this length would match the width of the weaving machine. POF can be integrated into any woven structure in the warp and weft directions. However, considering that POF should be connected to light sources in order to transmit light signal over a long distance, it is desirable for POF to be used without disconnection in a textile structure.

The bending condition applied to POF during the weaving process is the major issue. A bend formed with a small radius of curvature increases mechanical tension and thus may result in the breaking of POF. Thus, critical bending parameters have to be defined before POF is applied in the process of weaving.

Some different weaving processes, which were developed by a Korean company, Mikwang Textiles Ltd., are used for preparing a prototype of the POF-woven fabric and briefly introduced here. The same PMMA fiber and polyester yarn as used in Figure 5.2 are used as weft and warp, respectively.

In the first approach, POF is partially inserted into the plain woven structure as weft yarn, as shown in Figure 5.5. In this design, the excessive number of POF ends may render it impractical for use in clothing manufacturing.

Second, a continuous POF-feeding method, which is aimed to solve the problem of partially inserted POF, is tried, as shown in Figure 5.6. POF is continuously interwoven as weft yarns in this design.

As mentioned above, the bending of POF is very important in the light transmission along the fiber. In the continuously interwoven POF fabric, POF is bent perpendicularly at the edges and thus causes the significant optical loss. To alleviate the problem the angles of POF at the edges were curved using a Jacquard machine, as shown in Figure 5.7. Figure 5.8 shows a smart shirt that is made with the POF fabric obtained from the last method. The optical loss of the shirt is about −40 dB. The

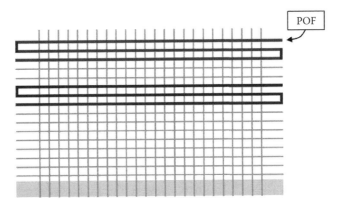

FIGURE 5.5 Schematic diagram of POF-inserted woven fabric.

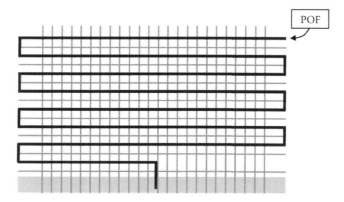

FIGURE 5.6   Schematic diagram of continuously interwoven POF fabric.

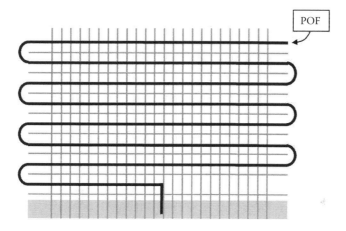

FIGURE 5.7   Schematic diagram of curved POF fabric.

major part of the loss occurs at curved edges, indicating that the control of bending loss is difficult, especially for such small diameter POF of 0.30 mm.

Since POF is usually supplied on spool, we should wind POF strand on bobbin before woven into fabric. Conventional fiber-winding method causes the twisting and bending of POF and even scratching on the surface. A shuttle loom machine was modified by adding a yarn-insertion device shown in Figure 5.9. It can prevent extra twisting and bending of POF so that the POF-integrated fabric can be manu-factured without breaking the fiber. Figure 5.10 shows a Jacquard loom machine with power-transmission device, which is located in a shuttle and prevents curved POF from getting loose in the woven structure.

5.3.1.2   Knitted Structure

Knitting is the interlooping of yarns to form a material. Knitted loop is usually referred to as stitches when they are pulled through another loop. There are two main classifications of knitted fabrics: weft knits and warp knits. In a knitted fabric, the

FIGURE 5.8   Smart shirt.

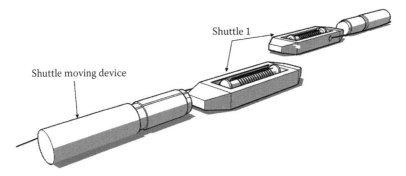

FIGURE 5.9   Weft yarn insertion device.

yarns are in the form of interloops that are exposed to severe bends at a very small radius of curvature. In this way, POF cannot transmit optical signal to a place that is far away. Therefore, POF needs to be integrated in a straight line, interlacing with the loops in knitted fabric. Weft or warp knitting is available. Figure 5.11 shows a POF-integrated, weft knitted fabric with a gauge of 11 in a continuous way.

5.3.1.3   Surface Treatment of POF Fabrics

As shown in Figure 5.2, POF becomes weak when some defects are introduced on its surface to impart the side-illuminating effect. This makes it impossible to manufacture the fabric using the artificially surface-defected fiber. An alternative approach is to

(a)

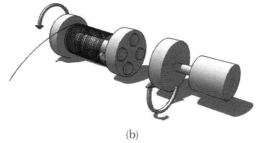

(b)

FIGURE 5.10 (a) Jacquard loom machine, (b) power-transmission device.

treat the surface of POF fabric, which is made of neat POF. Plasma treatment has been increasingly used for the surface modification of polymers without affecting their bulk properties (d'Agostino et al. 2003). Plasma is a cluster of particles including equal numbers of positive ions and electrons, free radicals, and natural species created by exciting a gas in electromagnetic or electrical fields. Low temperature plasma is generally created by advancing the process gas, such as argon or oxygen, into an excited state by means of an alternating current at radio frequency applied at a relatively low pressure.

The principal changes brought about by exposure of a polymer to plasma are in the surface wettability, the molecular weight of a surface layer, and the chemical composition of the surface. It is known that the effects of the plasma treatment are confined to a surface layer, 1–10 μm in depth.

The scanning electron microscopy (SEM) micrographs of POF treated with low temperature plasma at different process times are shown in Figure 5.12. A closer SEM view reveals some changes on the POF surface. Cracking to a certain extent was observed in the modified surface, especially on the POF surface treated for a process time of 30 min, which is supposed to make POF shine dimly. The details on the treatment conditions are described in Table 5.2.

FIGURE 5.11   POF-integrated weft knitted fabric.

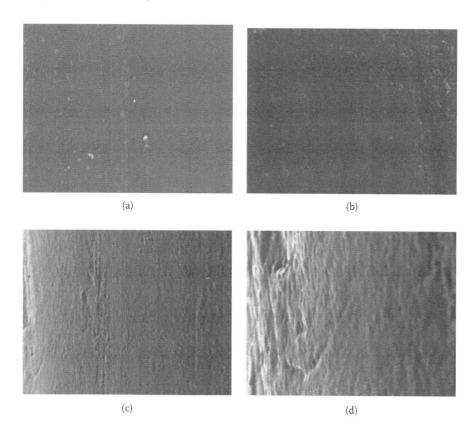

(a) (b)

(c) (d)

FIGURE 5.12 SEM image of POF (a) control, (b) plasma treated for 10 min, (c) plasma treated for 20 min, (d) plasma treated for 30 min.

TABLE 5.2
Plasma Treatment Condition

Gas	O₂/Ar
Gas flow rate (sccm)	1,300
Working pressure (mTorr)	60
Power (W)	2,000
Time (min)	10, 20, and 30

5.4   APPLICATIONS FOR POF-WOVEN FABRICS

According to the role of POF in woven fabrics, they can be categorized into two classes: optical signal transmission textiles, sometimes referred to as smart shirts or wearable motherboard, and photonic textiles. Here, we introduced some examples of POF-woven fabric applications. More detailed information about them is easily found on the Web.

5.4.1   SMART SHIRTS

The smart shirt made of POF-woven fabric is a flexible, wearable open platform that can be customized to monitor vital signs, external impacts, and other data through sensors woven into its fabric (Park, Mackenzie, and Jayaraman 2002; Sensatex 2005). Electrical and optical conductive fibers are woven or knitted with conventional textile fibers such as polyester or cotton and connected to the data bus and then to the main processor. The smart shirt functions like a motherboard and the POFs being integrated only act as an optical signal transmission channel. Sensors integrated into the shirt track vital signs such as heart rate, body temperature, and respiration rate. The role of POF in smart shirts is not large compared to electrical conductive fibers, which can be connected to various types of sensors based on the electronics. The lack of reliability of POF is also a problem. Although the shirt was originally designed for soldiers in combat, it is expected to be used for medical monitoring, clinical trials monitoring, athletics, and biofeedback. Because of its capability to monitor these vital signs, the shirt is being marketed as a way to prevent sudden infant death syndrome (SIDS).

5.4.2   LIGHT-ILLUMINATING TEXTILES

The idea of weaving side-emitting fiber into fabric was realized by Luminex in Italy and chosen as one of the "coolest inventions" of the year by *TIME* magazine in 2003. Light is emitted from the defects on the surface of POF, which seem to be created by mechanical notching and are viewed by SEM. The woven side-emitting POFs are end-irradiated at various colors by high-efficiency LEDs so that the wearer can really light up a room when they enter it (Luminex 2003). The power comes from an ordinary battery and varies according to the function. The fabrics are being used in stage costumes, handbags, and curtains, as well as clothing. Some samples are shown in Figure 5.13.

Woven POF fabrics can also be built up into a panel or other lighting device for membrane switching and surgical illuminator (Lumitex 2005). The device consists of one or multiple layers of woven POF fabrics placed between two different laminates such as back reflector and diffuser. Weaving technology to make uniform light distribution determines the performance of the devices.

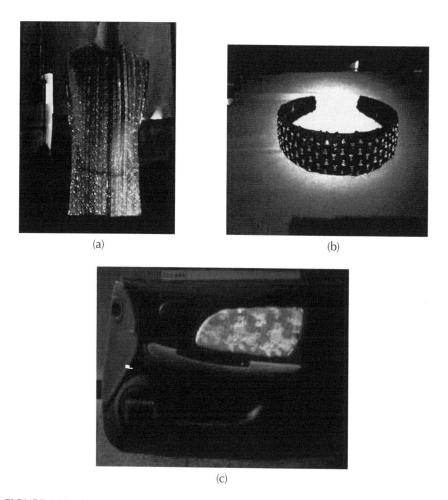

(a)

(b)

(c)

FIGURE 5.13 Some typical examples for fashion items using side-illuminating textiles. (a) Vest, (b) hair band, (c) door trim.

5.4.3 Flexible Display

Researchers at France Telecom have developed a prototype for a flexible screen made of woven OFs capable of downloading and displaying animated graphics on the wearer. Using a traditional two-dimensional loom, the OFs with a diameter of 0.25 mm or 0.5 mm are woven or placed in a chain. The pixels of the screen are directly formed on OFs, which have micro-perforation to emit light on their surfaces. Two different approaches, sandblasting and solvent spraying, are attempted to create the light-illuminating zone. The electronic components including LEDs are soldered on a flexible circuit board and then packaged in a fabric-layered sandwich. This offers an optimized display while maintaining very good flexibility and a comfortable yet resistant textile feeling. Although initially developed for clothing, they could be used in cars, portable electronic devices, and even houses. The details are well described in a reference (Koncar 2005).

5.4.4 Cure and Health Monitoring

POF-woven fabrics capable of emitting light can be used for light therapy, which has proven effective in treating acne vulgaris, delayed sleep phase syndrome, non-seasonal depression, psoriasis, and pain. It consists of exposure to daylight or to specific wavelengths of light. Red and near-infrared light are in a "window" of wavelengths that are able to pass through tissue up to 1 inch deep. The ideal wavelengths are between 600 and 900 nm, with the best results at specific ranges: 610–625, 660–690, 750–770, and 815–860 nm (Heelspurs 2006).

A flexible woven fabric that emits a blue light having a wavelength between 390 and 475 nm, called a Bili-Blanket, has been developed to treat jaundice in newborn babies by Lumitex. The blanket consists of two layers of parallel PMMA fibers, which are woven together with polyester cross fibers. The ends of the OFs are gathered together at one side and plugged into a light source. The product is expected to be helpful for newborn babies with jaundice because they can receive the treatment while being cuddled and fed, by wrapping them in a light-emitting blanket.

Researchers in Switzerland have published a proof-of-principle study demonstrating that POFs can be woven or sewn into textiles to create wearable health-monitoring devices. The team created a pulse oximeter in the form of a glove, where POFs were integrated into the fabric (Rothmaier, Selm, Spichtig et al. 2008). The results confirm the feasibility of textile-based oximetry and its promising potential as a wearable health-monitoring device. Wearable devices, such as a pulse oximeter, offer the promise of continuous and autonomous monitoring of vital health indicators. In the future, the team plans to integrate external electronic and optic devices into the textile using a small printed circuit board with diodes and phototransistors.

5.4.5 Textile Sensors

POF-woven textiles can be used to measure the changes in physical dimensions such as pressure and strain applied to the textile. Unlike most of the sensors based on microelectronics or conductive yarn, researchers from Switzerland have developed pressure sensitive textile prototypes based on POF-woven textiles (Rothmaier, Phi Luong, and Clemens 2008). Considering the flexibility, which determines the sensitivity of the pressure sensor, thermoplastic silicone polymers with high transparency and fully reversible elastic deformation are used instead of typical PMMA fibers. The sensor is designed to measure a change in transmitted light intensity caused by the deformation, which occurs when pressure is applied to the textile. It is known that its working range is between 0 and 30 N with a small drift in the range of 0.2% to 4.6%. The research efforts focused on developing optical sensors using POF-woven fabrics are relatively low compared to those of using POF itself.

5.5 PROSPECTS

POFs are very attractive in the development of new technologies to successfully miniaturize and embed electronics, optics, and sensors into fabrics and garments. They afford some advantages such as comfort and ease of movement for the wearer and immunity to electromagnetic interference. However, they play a minor role in transferring signals through the data link integrated into fabric compared to electric wire or conductive yarn. This is mainly due to their lack of reliability and durability when functioning as the data transmission channel, especially for bare POF. Although conventional textile machines can be used, abrasion of the POF surface during fabric manufacturing and macro-bending of the POF itself during integration into the textile structure are both unavoidable. They cause extra optical loss and weakening of the POF in fabric. Another limiting factor is the devices connected to OFs, most of which are designed based on microelectronics, not optics. POF-woven fabrics need additional and unnecessary optoelectronic devices, which convert the electrical signals to corresponding optical signals, for POF interactive fabrics. Such devices are not necessary for electrical interactive fabrics.

Although the future of POF-woven textiles for data signal transmission remains doubtful, it is very promising for various applications using the light-illuminating effect of POF, such as photonic textiles. A two-dimensional, large-area, flexible lighting system can be realized using side-emitting POFs. Many types of product, including prototypes, can be easily found on the Internet. Such products are aimed at applications ranging from consumer textile products, such as curtains and tablecloths, to flexible fabric screens and health monitoring sensors. The technologies for making side-emitting POF and then weaving without fiber breakage are the keys to success. Computer-controlled weaving technology, which can create optimized microbends in the fabric to produce a uniform light distribution, is a possible solution. Many components, including light source and battery, are required to operate photonic textiles. Except for some components presently available in the market, others, especially connectors and adaptors, need to be independently developed according to the applications of photonic textiles. Standardized components will help minimize such efforts and push photonic textile technology into the mainstream. It is believed that the medical market as well as the fashion industry will drive the technology forward and ultimately utilize its most important applications.

ACKNOWLEDGMENTS

This work was financially supported by the Center for Photonic Materials and Devices at Chonnam National University, and by a grant (10016447) from the Ministry of Knowledge Economy, Republic of Korea.

E. J. Park appreciates the financial support from the Center for Functional Nano Fine Chemicals, one of the BK21 programs from the Ministry of Education, Science and Technology, Republic of Korea.

REFERENCES

Asahi-Kasei. 2008. http://www.asahi-kasei.co.jp/pof/en/index.html.
Asahi Glass. 2008. http://www.agc.co.jp/english/rd_e/e_lucina1.html.
Atsuji, M., M. Tetsuhiko, H. Keiichi, and Y. Shinji. 2006. Optical properties of woven fabrics by plastic optical fiber. *J Tex Eng* 52:93–97.

Clemens, F., M. Wegmann, T. Graule, A. Mathewson, T. Healy, J. Donnelly, A. Ullsperger, W. Hartmann, and C. Papadas. 2003. Computing fibers: A novel fiber for intelligent fabrics? *Adv Eng Mater* 5:682–87.

d'Agostino, R., P. Favia, C. Oehr, and M. R. Wertheimer. 2003. *Plasma processes and polymer: 16th international symposium on plasma chemistry*. New York: Wiley Intersciences.

Daum, W., J. Krauser, P. J. Zamzow, and O. Ziemann. 2002. *Polymer optical fibers for data communication*. Berlin: Springer, 389–91.

Endruweit, A., A. C. Long, and M. S. Johnson. 2008. Textile composites with integrated optical fibres: Quantification of the influence of single and multiple fibre bends on the light transmission using a Monte Carlo ray-tracing method. *Smart Materials and Structures* 17:15004.

Graham-Rowe, D. 2006. Photonic fabrics take shape. *Nature Photonics* 1:6–7.

Harlin A., M. Mäkinen, and A. Vuorivirta. 2002. Development of polymeric optical fibre fabrics as illumination elements and textile displays. *AUTEX Research Journal* 3:1–8.

Heelspurs. 2006. http://heelspurs.com/led.html.

Im, M. H., E. J. Park, C. H. Kim, and M. S. Lee. 2007. Modification of plastic optical fiber for side-illumination. *Human-Computer Interaction. Interaction Platforms and Techniques* 1123–29.

Koncar, V. 2005. Optical fiber fabric displays. *Optics and Photonic News* 40–44.

Lasermet Ltd. 2008. http://www.lasermet.com/resources/classification_overview.htm.

Lekishvili, O., L. Nadareishvili, G. Zaikov, and L. Khananashvili. 2002. *Polymers and polymeric materials for fiber and gradient optics*. Utrecht: VSP.

Luminex. 2003. http://www.luminex.it/.

Lumitex. 2005. http://www.lumitex.com/.

Marcoe, J. (Ed). 1997. *Plastic optical fibers: Practical applications*. New York: John Wiley and Sons.

Mitsubishi Rayon. 2003. http://www.pofeska.com/pofeskae/pofeskae.htm.

Park, S., K. Mackenzie, and S. Jayaraman. 2002. The wearable motherboard: A framework for Personalized Mobile Information Processing (PMIP). *Proceedings of the 39th Conference on Design Automation*. ACM Press, 170–74.

Philips. 2008. http://www.research.philips.com/newscenter/archive/2005/050902-phottext.html.

Rothmaier, M., M. Phi Luong, and F. Clemens. 2008. Textile pressure sensor made of flexible plastic optical fibers. *Sensors* 8:4318–29.

Rothmaier, M., B. Selm, S. Spichtig, D. Haensse, and M. Wolf. 2008. Photonic textiles for pulse oximetry. *Optics Express* 16:12973–86.

Sansatex. 2005. http://www.sensatex.com/.

Spigulis, J. 2005. Side-emitting fibers brighten our world. *Optics and Photonics News* 16:34–39.

Toray. 2008. http://www.toray.co.jp/english/raytela/index.html.

Ziemann, O., H. Poisel, M. Luber, M. Bloos, and A. Bachmann. 2005. Optical data transmission for short distances—technologies and opportunities *Lasers and Photonic* 4:36–41.

6 Hardware and Software Architectures for Electronic Textiles

Mark T. Jones and Thomas L. Martin

CONTENTS

6.1   INTRODUCTION

Electronic textiles (e-textiles) are fabrics where computing elements, sensors, actuators, and networks are an intrinsic part of the cloth. E-textiles are an enabling technology for wearable computers (Starner 2001a,b) that look like normal clothing and for intelligent room furnishings in a pervasive computing environment. In this chapter, we make the case that e-textiles occupy a unique corner of the distributed, embedded computing design space, and therefore require a hardware and software architecture that is tailored to this space. While previously reported research in e-textiles has primarily examined individual applications, the technology itself supports, and to a large extent requires, a computing architecture that simultaneously supports multiple applications and heterogeneous sensors and computing elements. For example, users should not be expected to choose between garments that support MP3 capability (Jung, Lauterbach, and Weber 2002), heart monitoring (Paradiso et al. 2003), and gait analysis (Edmison et al. 2003); all three capabilities and more should be simultaneously and reliably available to the user. To draw an analogy with desktop computing, the current state of the art of e-textiles is equivalent to requiring a different desktop computer for each application that the user would like to run: one computer for word processing, one computer for using a spreadsheet, and a third for browsing the Web. Instead, a single e-textile fabric should support multiple applications running simultaneously, each with a range of sensing and computation requirements.

In addition to the promise of such applications, advantages claimed for e-textiles over the hard-shelled, non-fabric form factors typical of today's consumer electronics (Marculescu et al. 2003) include inherent fault tolerance through the availability of many fibers (e.g., multiple power distribution busses), improved application performance due to the capability to deploy many sensors (Shenoy 2003), low cost due to the use of high-volume textile manufacturing techniques, robust power-aware operation due to the availability of many fibers (e.g., communication busses) (Jones et al. 2003), and a greater likelihood of user acceptance due to the familiarity of users with textiles. Reaping these advantages, however, requires, among other things, the support of a computer architecture, including hardware and software, tailored to this environment.

In this chapter, we lay out the case for the need for such an architecture and then propose a two-tier hardware architecture and a publish-subscribe software architecture to meet this need. The hardware aspect of the architecture is designed to provide fault-tolerant, power-aware, wired communication between many on-fabric nodes. The architecture avoids the use of paths of any significant length in which sensor data is communicated in analog form; such communication paths have inherent weaknesses, for example, in the areas of fault-tolerance and susceptibility to noise. To facilitate this network, while ensuring that the architecture maintains the properties of low cost and small physical size, the first tier of the hardware architecture is composed of very small, low-cost computing devices that are paired with sensors. The second tier is composed of a small number of more powerful computing devices that are capable of running application algorithms, for example, an application that classifies the user's activity such as that described in (Jolly 2006).

Without an appropriate software architecture, however, this hardware architecture would be difficult for application developers to program and would receive no acceptance in a wider community. To address this need, we propose a hierarchical, services-based architecture that allows an application developer to program a garment instead of a heterogeneous collection of computing devices. This software architecture is designed such that very simple services (e.g., producing a stream of accelerometer data) can run on the first-tier nodes, while allowing applications to draw on more complex services that include the discovery of sensor types and locations, location-based addressing, and fault-tolerance communication. We have designed the system using a combination of simulation and implementation. In addition to using simulation to design the system, we have designed the simulation environment as an emulation environment as well, i.e., a developer can write a program in the simulation environment and move that program to a physical implementation of the architecture as described in (Zeh 2006) and (Graumann et al. 2007).

The remainder of this chapter is organized as follows: Section 2 gives a brief overview of the properties of e-textiles as they are relevant to this chapter. The case for digital, wired in-garment communication is made in Section 3. Section 3.2 outlines the proposed hardware architecture and Section 4 lays out the corresponding software architecture, with concluding remarks in Section 5.

6.2 PROPERTIES OF E-TEXTILES

Papers such as (Lind et al. 1997), (Nakad et al. 2007), and (Marculescu et al. 2003) have described the properties of e-textiles across a range of applications. These properties are briefly summarized in this section, with a focus on wearable e-textile garments, to provide the basis for the analyses in the following sections.

6.2.1 WEARABILITY

Garments are inherently easy to use; after a certain age, the vast majority of our population is capable of dressing themselves. E-textile applications can capitalize on this familiarity to simplify the deployment of a computing/sensing system. For example, a pair of pants designed for gait analysis will, by default, have sensors correctly positioned if the wearer has selected a garment of the correct size. While correct positioning is generally the default situation, there are exceptions; for example, a wearer may roll up the sleeves of a shirt. In contrast to a garment-based system, a system in which discrete devices are attached to the body, for example, motion capture system reflective markers and body-mounted wireless sensor nodes, requires a significant period of time to correctly position (Edmison et al. 2003). If time and care are not taken to correctly position the devices in such systems, the discrete devices will have a large variability in placement each time they are put on, which must be accounted for in the day-to-day use of the system. Consequently, garment-based e-textile systems have the advantage of knowing that the position of the sensors is fixed on the garment.

Another advantage of garments is that they do not encounter the user acceptance barriers of currently available hard-shelled electronic devices, assuming, of course, that the garments meet fashion and comfort expectations of the wearers. Systems in new and unfamiliar shapes, sizes, or materials generally encounter resistance from many users. For example, fall-mitigation devices, even among those prone to falling, are not widely accepted (Edmison et al. 2003). To "coat-tail" on the acceptance of garments, an e-textile system must not alter the existing garment form factor. Eventually, it may be possible to change the public's fashion sense as they come to understand the benefits of an intelligent garment, analogous to the acceptance (rather surprising to the authors) of Bluetooth headsets for cell phones. This would permit clothing designers to capitalize on aspects of the electronic components to create new fashions.

In the meantime, assuming that an e-textile garment must look like a normal garment implies that the additional components must either look like traditional garment components, be hidden, and/or be very small. Garment components such as buttons, sequins, zippers, and rivets offer opportunities for the integration and attachment of new components such as integrated circuits and discrete sensors (Lehn et al. 2004). Integral aspects of many garments, such as cuffs, collars, and seams, offer locations in which to hide similar components. Direct connection and/or integration into the garment is acceptable as long as the components are small and/or flexible enough so as not to be noticeable to the eye or an irritant to the wearer. As e-textiles grow in popularity, it will be desirable to have electronic components become available in form factors that are more compatible with textile- and garment-making techniques. For example, it will be advantageous to have components spun into fiber form factors instead of attaching them as on printed circuit boards in their currently available standard electronic packaging. Fiber form factor components can be woven directly into the fabric without the need for attaching them as discrete components. While the flexible circuit boards and cloth circuit boards such as those described in (Buechley 2006) are steps in the right direction, they are far from ideal.

To be feasible for mass production, e-textile applications must work for the wearer without extensive tailoring and, ideally, the application will work with no tailoring at all. It is unlikely that a simple, "one size fits all" approach will be successful because traditional garment designers draw upon a vast range of components to meet utility, fashion, and comfort requirements across a range of body sizes for both adults and children. If, however, a person wears a size "42" suit jacket, he should be able to take a size "42" e-textile suit jacket off the rack, put it on, and have the garment's applications function correctly without tailoring.

Aside from the expectations described above, there are additional expectations held by most garment wearers that must be met if we are to capitalize on the wearer's acceptance of this form factor. The first of these is cost; for the majority of applications and the majority of wearers, if the addition of "e-textile" functionality significantly increases the cost of the garment relative to a non-intelligent version of the garment, then there will be an additional barrier to acceptance. While costs will vary significantly depending upon the application and the technology (i.e., the electronics) employed, the analysis in (Graumann et al. 2007) suggests that it is possible to keep costs in line with traditional textile costs.

6.2.2 RELIABILITY

In addition to cost, wearers will expect garments to continue to be useful even in the presence of small flaws due to daily "wear and tear." For example, most wearers do not notice when a single fiber in a garment becomes very worn or even broken, and certainly do not discard the garment due to such a flaw. An e-textile should not cease to function when similar flaws are present; in fact, the large surface area and number of fibers offer an opportunity for extensive fault tolerance.

One particular issue for e-textile garments is that a short-circuit anywhere in the fabric could potentially rapidly exhaust the battery. This precludes the use of uninsulated conductive fibers for garments where the fabric can fold back on itself, and is one of the motivating factors for our choice of weaving insulated wires into the fabric for the power and digital communication network of our e-textiles (Lehn et al. 2004). One possibility for withstanding short circuits in the event of a flaw is to use multiple batteries for different sections of the fabric; our previous work on large-scale (non-wearable) fabrics has shown this to be an effective way to maintain the performance of an e-textile even in the presence of multiple short-circuits (Martin et al. 2004; Sheikh 2003). However, using multiple batteries in a garment will result in the wearer having to replace or recharge the batteries more often than if all of the energy were contained in a single battery, unless some provision is made for sharing energy between batteries. But this mechanism will itself have to be robust in the presence of short circuits.*

A final aspect of reliability is that the physical form factor of e-textiles, with long parallel runs of conductors, may result in electrical noise problems on signals, particular for analog signals. This is another motivating factor behind our digital communication network in the fabric: each sensor is co-located with a microcontroller that converts the sensor's analog signal to digital and then puts the digital data on the network, which is much less susceptible to noise. Section 3 will describe this in greater detail. Furthermore, any noise induced on the sensor itself can be filtered locally by the microcontroller before the data is transferred. For example, the rug described in (Graumann et al. 2007) had long parallel runs of electroluminescent (EL) wire and piezoelectric cable. The EL wire ran at a high voltage (approximately 100 V) and frequency (4 kHz) that coupled noise into the piezoelectric cable; this noise was removed with a digital filter running on the processing node for the piezoelectric cable before being sent out onto the network.

6.2.3 ADAPTIVE FUNCTIONALITY

E-textile garments will ultimately need to flexibly accommodate a wide range of applications, with particular combinations tailored to the needs and preferences of individual wearers. Even at this nascent stage in the development of e-textiles, the range of applications includes health (Park et al. 1999), entertainment (Jung, Lauterbach, and Weber 2002), and building navigation (Chandra, Jones, and Martin 2004). The number of applications is likely to exceed the number of garments that a user is willing to wear at any given time. Thus, it is ultimately undesirable to have single function garments: a user would be forced to choose, for example, between a garment to monitor heart/respiration activity and a garment to monitor physical activity. Garments that can accommodate multiple applications, however, must ensure that resources (such as energy, network bandwidth, and computation) are allocated according to the relative importance of the applications, where the relative importance may be defined at design time or by the wearer, and it may vary with time and conditions.

Each application is likely to require different types of sensors and actuators, as well as a range of computational performance. Despite these differing requirements, it is desirable for a garment to allow new applications to be added without causing resource conflicts or existing applications to be reconfigured. Furthermore, adding new components should not require a redesign of the garment.

---

* We leave puns on short circuits in e-textile garments as an exercise for the reader. However, "shirt circuits" and "shorts in my shorts" spring quickly to mind.

6.3 A HARDWARE ARCHITECTURE FOR ELECTRONIC TEXTILES

This section describes our hardware architecture for e-textiles, designed with the properties of Section 2 in mind. This section first makes the case for a wired, on-fabric digital communication network, and then presents a two-tier system of heterogeneous processing nodes.

6.3.1 ON-FABRIC DIGITAL COMMUNICATION NETWORK

The majority of e-textiles developed to date have relied primarily on a communication architecture in which a collection of sensors distributed across a garment are connected to a single collection point via analog conductive fibers, e.g., (Buechley 2006; Mattmann et al. 2007). Such an approach is suitable for prototyping e-textile technology and applications, but it ultimately presents unacceptable barriers to the uses envisioned for e-textiles, as analyzed below.

- The communication of analog sensor data typically requires dedicated communication paths, precluding the type of expansible system required for executing a changing mix of *multiple applications*. In contrast, a digital, switched network can multiplex data from many sources.
- Ensuring *signal integrity* along relatively long paths is often difficult for analog sensor data and is typically application dependent. In contrast, signal integrity of a digital, switched network can be addressed in the design process one time and re-used because it is independent of the sensor data.
- Each new sensor requires its own discrete path, *complicating the design process* by forcing a redesign of the physical textile for different sensor configurations. In contrast, a digital, switched network can accommodate new nodes without any redesign until the bandwidth required by the nodes exceeds the effective bandwidth of the system. This limit on the bandwidth can be mitigated with proper design of the network.
- The central processing device(s) must be able to accommodate potentially large numbers of sensor nodes. The number of input/output connections to the fabric will increase with the number of sensors, presenting *packaging and attachment* issues that will increase costs. In contrast, a digital, switched network allows for a fixed number of input/output lines for each network interface.
- The *fault tolerance* of the system is reduced because each sensor requires at least one continuous path from the sensor to the processing unit. In contrast, a digital, switched network can route around points of failure in the garment.

A wire-based network offers significant advantages, as analyzed below, over a wireless network for the components on an e-textile. While such a conclusion may seem obvious in light of present-day computing technology, it is not widely accepted in the fields of wearable computing and body sensor networks, particularly body sensor networks for medical applications. The analysis below is directed towards the primary communication network on the e-textiles, but the analysis does not apply to off-fabric communication, which is better handled by wireless communication: there is typically a need for the garments to communicate with another garment or with the surrounding infrastructure without being physically tethered either to the other garment or to the infrastructure. For example, wireless communications can be used to link an e-textile shirt and pair of pants or an undergarment and a jacket, to allow a health monitoring application to periodically report results to a doctor's office or to enable an MP3 player to download new music. These off-fabric communications are better suited to one of the readily available wireless technologies, e.g., WiFi, Bluetooth, or Zigbee, depending upon the data bandwidth and distance required for the application. Our e-textile pants (Edmison et al. 2003) and jumpsuit (Chong 2008) both use a Bluetooth link to communicate with a nearby laptop, but all communication between sensors on the garments is carried on a wired network woven into the fabric.

While communications off-fabric are wireless, for communication between the elements on the e-textile, a wire-based network is superior in many respects. In the following paragraphs, several aspects of a wearable system are considered, including energy storage and distribution, system deployment, electro-magnetic emission, and low-power operation.

The type of network chosen has a significant effect on how wearable systems are deployed and operated, particularly for the types of applications mentioned in previous sections, in which (at least) tens of sensors are distributed around a user's body. A wireless network initially appears to offer many of the same deployment advantages that are present in wireless home networks as compared to wired home networks. For example, a user can simply fasten (e.g., by Velcro) wireless sensor nodes to his/her body and have them quickly participating in a network. This approach, however, requires the user to spend the time to attach the sensor nodes; further, it requires the user to attach the correct sensors to the correct locations on the body. In comparison, an e-textile system, in which the garment itself contains the conductive fibers necessary for communication, requires a user to simply put on a garment(s). The e-textile approach is not without cost because it does not typically allow for sensors or computing nodes to be moved from garment to garment; such re-use of nodes across garments is possible, but it gives up some of the advantages described above.*

The operation of a wearable computing system is typically faced with a severe constraint on the amount of energy that can be stored or harvested during operation. Batteries are typically used for energy storage, and although battery form factors are becoming more compatible with wearable systems, the energy density of batteries has remained largely constant (Paradiso and Starner 2005). Wearable systems must typically track battery status and allow for batteries to either be replaced or recharged. This presents a drawback for wireless approaches because each sensor node and/or computing node (or local groups of nodes) requires its own battery. Requiring a user to manage tens of batteries is an unrealistic approach. Assuming there is some variation in the power consumption of the wireless nodes, the batteries of the nodes will not all be exhausted at the same time. Consequently, the user

---

* With the exception of some types of sensors, e-textiles are compatible with typical garment laundering processes, e.g., Buechley (2006).

will have to change or recharge each battery as it becomes exhausted. In contrast, an e-textile garment is effectively a backplane for power as well as communication, allowing for the number and location of batteries to be independent of the location of the sensing and computing nodes. This system avoids the unnecessary duplication of battery packaging and monitoring mechanisms present in a wireless system.

There are several concerns with respect to the electromagnetic profile of a wearable computing system. User concerns regarding the security of their personal data in computers and computer networks continue to be well founded. Wearable computing systems have the potential to generate more data that is even more personal, raising this concern even higher. A wireless network for a wearable network must address such security concerns without imposing a significant burden on the wearer; this includes not only during operation, but also during deployment. However such security is provided, it imposes costs in terms of hardware, software, network efficiency, and energy consumption. In contrast, an e-textile system can typically rely upon on the physical security of the conductive fibers in the garment and transmit "clear text" within the garment. Both systems, of course, have in common security issues that must be addressed such as authenticating the wearer and ensuring the security of information stored on the system.

In addition to security, many people are concerned with the health effects of radio frequency (RF) devices. Whether or not these concerns are well founded, systems that extensively use RF communication in proximity to the body face a barrier to adoption. To a similar, but lesser extent, any wearable computing system faces user concerns about any electromagnetic emissions in proximity to the body. Beyond the health effects of RF, there are also data transmission losses that can result from, for example, interfering RF emissions from other devices. Concerns about the potential for interference may also limit the use of wireless sensors in certain locations, for example, in hospitals and airplanes, or impose costly regulatory requirements on their design.

Perhaps the most fundamental issue, however, is that of energy consumption, for which an e-textile solution offers both obvious and less obvious advantages. Wireless transmission and reception of data clearly requires significantly more energy than wired transmission and reception of data over the same distance. A typical body-worn sensor, however, has a low communication rate, even when reporting data without any processing and/or filtering. For example, an accelerometer in a gait-monitoring application may need to report on the order of a hundred one-byte samples per second (Edmison et al. 2003), a fraction of the bandwidth available in wireless networks. To examine and put into context the energy required to send and receive such data, we consider two commercial devices: a microcontroller and a Zigbee transceiver. The current consumption numbers for these devices in various operating states are given in Table 6.1. Assuming perfect efficiency at a transmission rate of 250 Kbps, the Zigbee device would require 0.232 mW to transmit the accelerometer data in the example above. The receiving unit would consume a comparable amount of energy, assuming it was only active when the transmitter was active.

On the surface, this is a reasonable power consumption rate as compared to the rates of the microcontroller in Table 6.1. This calculation is not realistic, however, because of the need to manage the entire system of sensor and computing nodes,

TABLE 6.1

The Current Consumption in Different Modes for a Microcontroller and a Low-Power Transceiver

Device	Operational Mode	Current Consumption (mA)
PIC1845J10	Sleep	0.025
PIC1845J10	Run @ 4 MHz	4.4
PIC1845J10	Run @ 40 MHz	11.5
MRF24J40	Sleep	0.002
MRF24J40	Receive	18
MRF24J40	Transmit	22

Note: The microcontroller is a member of the well-known PIC family produced by Microchip. The transceiver is a Zigbee device also produced by Microchip. The results given are quoted as typical from the manufacturer's datasheets for both devices operating at 3.3 V.

not just an individual pair. It is reasonable to assume that a single node is the active receiver of sensor information; this node would have a high power consumption rate in comparison to most other nodes, as it would always be in the receive mode. The Zigbee device in Table 6.1 requires 59.4 mW in receive mode. A larger difficulty is the need to actively manage all of the nodes in the system in order to reduce the power consumption of the entire system. For example, during most of the day, the majority of the sensor nodes can be inactive in the gait-monitoring application because the user is not walking; a small number of nodes may be active and, when the user begins to move, these nodes can activate the remaining nodes. To reduce power consumption, it is desirable for those nodes to be in a sleep mode, including the computational and communication components of the node. In a sleep mode, the power consumption of both the transceiver and the microcontroller in Table 6.1 are in the microwatt range rather than the milliwatt range. To be useful, of course, these nodes must be able to be quickly activated. Such an activation, in a wireless network, can only take place if a node is in a receive mode when a sending node is transmitting the signal to awaken. By coordinating time slots, it is possible for a node to periodically reawaken to "check in" and reduce power consumption; such an approach does result in delays in activation. In contrast, microcontrollers in a sleep mode can be reawakened nearly instantaneously via signaling on a wired network.

6.3.2 Two-Tier Processing Architecture

The previous sections laid out requirements that include digital communication from sensors, the capability to simultaneously execute multiple applications, small physical size, and a need to keep the costs comparable to traditional garments. Given these constraints, we have opted for a two-tier architecture in which there are a large number of small tier-one nodes (primarily sensor nodes) and a small number of more powerful tier-two nodes capable of executing application code. As will be described in Section 6.4, the tier-one nodes will provide data to the tier-two nodes, which will manage the tier-one nodes and run applications. The tier-one nodes are intended to sample sensor data, to perform minimal processing of sensor data (for example, low-complexity filtering or averaging), and to carry out simple networking; they are intended to be as small and as power-efficient as possible. Sampling rates are typically tied to macroscopic phenomena associated with the human body, for example, walking, respiration, and gestures, which can be measured adequately with sampling rates on the order of 1 Hz to 1 kHz. Because the sampling rate requirements are primarily associated with such phenomena and because tier-one nodes are expected to do only minimal processing and networking and leave the more complex application-level processing to the tier-two nodes, the requirements for tier-one nodes can be expected to remain the same across a wide range of applications and over time. As very large scale integrated (VLSI) technology improves, these tier-one nodes should become smaller and more power efficient instead of becoming more capable. Tier-two nodes, however, are primarily intended to manage the tier-one nodes and to run applications. Consequently, their requirements will likely change as applications become more demanding. While the requirements on the tier-one nodes drive them towards homogeneity, there is no such expectation for the tier-two nodes. The tier-two nodes will be heterogeneous with a range of performance requirements determined by the applications. In terms of numbers, for most applications there will be at least an order of magnitude more tier-one nodes than tier-two nodes.

To take advantage of the large number of fibers in an e-textile, the architecture has many redundant paths in the wired network. Both the nature of the fabric construction (e.g., weaving) as well as the nature of garment construction dictate that there are physically disjoint paths in the network that are connected by tier-two routing nodes. Because of their limited nature, the network demands placed upon the tier-one nodes are reduced by ensuring that each tier-one node is on the same physical network path as at least one tier-two node. It is the responsibility of these tier-two nodes to manage tier-one nodes.

Our current physical implementation of the architecture, taking advantage of commercially available processors and microcontrollers, is based upon 8-bit Atmel microcontrollers for the tier-one sensor nodes. Each of these microcontrollers has an analog-to-digital converter to read its respective sensor's output, and a digital serial port that is used to network it to the other nodes on the fabric, using a standard I^2C two-wire serial interface (NXP Semiconductors 2007). These tier-one sensor nodes are implemented as small printed circuit boards that attach directly to conductive fibers woven in both the warp and fill directions of the fabric, using insulation displacement connectors (Lehn et al. 2004).

Figure 6.1 shows a close-up of an e-textile rug (Graumann et al. 2007) with a tier-one node attached. The four conductive fibers of the network interface consist of two fibers for power and ground, and two fibers for the I^2C interface.

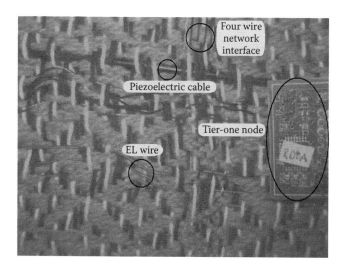

FIGURE 6.1   In this example, accelerometer sensor nodes are located on the limbs along with gyroscopic sensor nodes on the waist; services on these nodes provide data for a body motion classification application running on a tier-two node. Tier-one sensor nodes located around the heart provide data for a heart analysis application running on a tier-two node.

Tier-two nodes use an ARM7-based microprocessor and may have more than one interface so that they can be on multiple networks at the same time. The network on the fabric consists of multiple independent I^2C busses, or sub-networks, with tier-two nodes acting as routers that connect the sub-networks. On an individual sub-network, nodes address each other using I^2C addresses. For sending a message to a node on another sub-network, the message is sent to the router, which then forwards the message to the router for the sub-network where the recipient is connected. If the recipient's sub-network is unknown, then the router sends the message to all other routers in the system (flooding), and when the flooded message is received at the recipient's sub-network, that router responds so that the sending router will know which sub-network to forward the message to. This forwarding mechanism is used when matching service publishers and subscribers, described in the next section.

6.4   THE SOFTWARE ARCHITECTURE

Without a relatively simple abstraction, programming a distributed, heterogeneous, real-time, fault-tolerant architecture, with multiple simultaneously executing applications, is a very difficult task that far exceeds the budgets and designer capabilities for the envisioned e-textile applications. Further complicating matters is that for each garment design, there will be a corresponding set of garment sizes that are likely to have different numbers and placement of sensors; it is desirable that the software and its behavior be largely independent of small variations in the number and placement of sensors. Finally, the abstraction must take into account that the tier-one nodes in the proposed architecture do not have the resources to implement the software approaches that are often used to support abstractions.

6.4.1   E-TEXTILE SOFTWARE SERVICES

To address this need, we have designed an event-driven, services-based software architecture. The architecture is hierarchical in that services are expected to use the results of other services to accomplish their tasks. Further, it is the responsibility of the system to map the execution of services to specific nodes. When a service request results from another service, it does not have any knowledge of where that service is executing, it is simply provided with the results. In the two-level hardware architecture of Section 3.2, the tier-one nodes are capable of executing a fixed set of simple services, while the tier-two nodes can execute a range of services as well as determine the assignment and migration of the execution of services. An application (as well as the system software) is composed of services.

Consider the concrete example of the hardware and software architecture in the e-textile jumpsuit shown in Figure 6.2, in which each tier-one node provides a service that is capable of producing sensor results at a specified frequency. The heart-monitoring and physical activity services running on tier-two nodes request sensor data at specific sampling rates from specific locations on the body; the applications

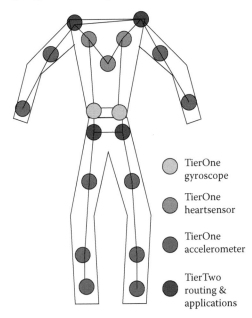

FIGURE 6.2   In this jumpsuit example, accelerometer sensor nodes are located on the limbs along with gyroscopic sensor nodes on the waist; services on these nodes provide data for a body motion classification application running on a tier-two node. Tier-one sensor nodes located around the heart provide data for a heart analysis application running on a tier-two node. Because the garment is a jumpsuit, wired connections between the torso and the legs are made in the seams. If this application were implemented as a separate shirt and pair of pants, then the connection between the torso and legs could be made either over a wireless connection or some wired connection that the user must connect after donning the garments.

do not request data from a specific node, instead the system determines which node(s) will be selected to meet each service request. For example, if accelerometer data is requested from the lower left leg, then the system will determine which sensors can provide that data (if any) and select the most appropriate.* This approach allows for the software to be written independently of the underlying hardware configuration and, in some cases, allows for faults to be tolerated (e.g., a sensor node providing a service can fail and another node with similar capabilities can replace it) without the application service being aware.

To meet the requirements of an e-textile system, as described in this chapter, the software system must have low overhead requirements and, in particular, must allow for participation by the very limited tier-one nodes. Because tier-one nodes are always on a sub-network with at least one tier-two node (as described in Section 3.2), the tier-one nodes must only be aware of how to communicate with a single node. It is the responsibility of the tier-two node to manage the interaction of "its" tier-one nodes with the rest of the system.† Because tier-one nodes are on the same network as all other nodes, they use the same network protocol as the rest of the system; however, they only need to respond to a subset of the protocol.

6.4.2   OVERVIEW OF THE PUBLISH–SUBSCRIBE ARCHITECTURE

The specific steps that the system follows to allow an application to find nodes providing required services is illustrated by Figure 6.3. First, immediately after powering on, each router (a tier-two node) sends a message to all of the tier-one nodes on its sub-network, requesting information about the services that they can provide, or *publish*. Then, each of the tier-one node publishers responds to its router's request, and the router creates a table of the services available on its sub-network (Figure 6.3a).

At some time after that, a node running an application may require a service to complete the application. That application node sends a request to its router that it would like to subscribe to a publisher of that service. Assuming that no publisher exists on that router's sub-network, the router floods the other routers with a request for the service (Figure 6.3b). If another router has a node on its sub-network with the requested service, it responds to the request with information about the address of the service. This response is sent to the application node that made the request (Figure 6.3c). If multiple publishers are available, the application node can select the publisher to which it would like to subscribe. The application node then sends a service subscription message to the router node managing the selected publisher; the router node notifies the publisher that it has a subscriber. Upon receiving the subscription request, the publisher begins providing the requested data to the application node (Figure 6.3d). As described in Section 3.2, this data is transmitted between the sub-networks by forwarding through the appropriate tier-two routers.

So far, this example has shown only how the publish-subscribe mechanism can be used to match applications with their required data, and for the sake of simplicity

---

* See Zeh (2006) for a discussion of the different bases upon which this selection can be made.
† To increase fault tolerance, the system allows for more than one tier-two node to manage a single tier-one node.

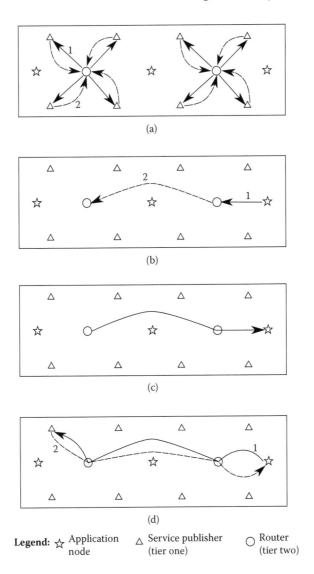

(a)

(b)

(c)

(d)

Legend: ☆ Application   △ Service publisher   ○ Router
           node   (tier one)   (tier two)

FIGURE 6.3 An illustration of the publish-subscribe mechanism used to connect nodes providing a service to application nodes requesting that service. (a) Routers request information on services from the tier-one nodes on their sub-network (*solid arrows, 1*). Tier-one nodes respond with available services (*dashed arrows, 2*). (b) An application node requests a service (*solid arrow, 1*), which is then flooded to other routers (*dashed arrow, 2*). (c) The router with that service responds to the service request. (d) The application node sends a service subscription request to the desired service publisher node (*solid arrow, 1*), which then begins providing the requested data to the application node (*dashed arrow, 2*).

the example has ignored the underlying network connections. The simplest network for implementing this example consists of each tier-one node being connected to only one network (to its respective tier-two router) and with each tier-two router node being connected to two networks (to its set of tier-one nodes and to the other router). However, using only this number of network connections does not provide any fault tolerance if there is damage to one of the networks.

To increase the level of fault tolerance, nodes must be given more network connections. For example, each of the tier-two routers can be placed on an additional network with some or all of the other routers such that, if one of a tier-two node's "router networks" is broken, it can still use the other router network to communicate with the sub-networks managed by the other tier-two nodes. This additional network connection will allow the fabric to continue to function if one of the router networks is broken, but it will not provide protection against one of the publisher/subscriber sub-networks failing. Tolerating a fault on one of those sub-networks requires additional network connections on each tier-one node, allowing it to have multiple connections to tier-two router nodes. The failure of any of the connections on either the tier-one or tier-two nodes may be used by higher-level applications to evaluate the damage to the garment and its level of functionality. However, the complexity and cost overhead of these additional network connections must be balanced against the e-textile application's reliability requirements and expected fault type and frequency.

Another level of fault tolerance can be provided by having several publishers of similar data, each on a different sub-network. In this configuration, a subscriber can choose between which one of the publishers to subscribe to, and in the event of a failure of one of those sub-networks, the subscriber can choose a different publisher. If not all of those publishers have data of the same level of quality, then the application can at least continue to function, albeit perhaps at reduced performance/accuracy.

6.5   CONCLUDING REMARKS

This chapter presented an argument for an approach to communication in e-textile systems. In particular, a case is made for digital, wired communication in e-textiles, with wireless communications to be used only for communicating off of the garment. This case is supported by observations regarding the construction and operation of e-textiles. This approach to communication on the e-textile is embodied in a two-tier hardware architecture outlined in the chapter. Applications running on this hardware architecture are created using a publish-subscribe software architecture that allows the application node to discover services that are available on the e-textile, allowing the work of the computation to be distributed throughout the e-textile.

In the near term, there is a need to investigate the performance of the system across a broader range of e-textile applications and to gather experimental results to guide design decisions made within the architecture. Should this approach be proven effective across a range of applications, there is a need for tier-one nodes and a network that is better suited to e-textiles. Even though the tier-one printed circuit boards can be made quite small and can be concealed in e-textiles, current sensors, microcontrollers, and other components are general purpose in nature and designed for inclusion in printed circuit boards; a design targeted to high-volume e-textile applications would be smaller and likely more power efficient. The I^2C network standard has not been designed for a large number of nodes nor has it been designed with interconnected I^2C busses. Further, because (at the time of this writing) the complexity of attaching an external device to fabric is related to the number of electrical connections being made, from the viewpoint of constructing e-textiles, the I^2C standard is less appropriate than, for example, the Dallas Semiconductor One-Wire network. However, communication on the One-Wire standard is too slow for our intended applications and is not supported by enough manufacturers for our prototyping work.

In the long term, it is our desire that hardware interfaces will be developed that are appropriate for e-textiles, particularly garments. We believe that the combination of our hardware/software architecture for e-textiles and interfaces that meet the requirements of wearability and manufacturability will allow e-textile applications that are currently only research prototypes to become widely adopted.

ACKNOWLEDGMENTS

An earlier version of this chapter originally appeared in (Jones, Martin, and Sawyer 2008). This material is based upon work supported by the National Science Foundation under Grant Nos. CCR-0219809, CNS-0447741, CNS-0454195, and CNS-0834490. Any opinions, findings, and conclusions or recommendations expressed in this material are those of the author(s) and do not necessarily reflect the views of the National Science Foundation.

REFERENCES

Buechley, L. 2006. A construction kit for electronic textiles. In *Proceedings of the Tenth International Symposium on Wearable Computers*, 83–90.

Chandra, M., M. Jones, and T. Martin. 2004. E-textiles for autonomous location awareness. In *Proceedings of the Eighth International Symposium on Wearable Computing*, 48–55.

Chong, J. 2008. Activity recognition processing in a self-contained wearable system. Master's thesis, Bradley Department of Electrical and Computing Engineering, Virginia Tech.

Edmison, J., M. Jones, T. Lockhart, and T. Martin. 2003. An e-textile system for motion analysis. In *International Workshop on New Generation of Wearable Systems for E-Health*, 215–23.

Graumann, D., M. Quirk, B. Sawyer, J. Chong, G. Raffa, M. Jones, and T. Martin. 2007. Large surface area electronic textiles for ubiquitous computing: A system approach. In *Mobiquitous 2007*. IEEE.

Jolly, V. 2006. Activity recognition using singular value decomposition. Master's thesis, Bradley Department of Electrical and Computing Engineering, Virginia Tech.

Jones, M., T. Martin, Z. Nakad, R. Shenoy, T. Sheikh, D. Lehn, and J. Edmison. 2003. Analyzing the use of e-textiles to improve application performance. In *Proceedings of the IEEE Vehicular Technology Conference 2003, Symposium on Wireless Ad Hoc, Sensor, and Wearable Networks*.

Jones, M., T. Martin, and B. Sawyer. 2008. An architecture for electronic textiles. In *Proceedings of the ICST Third International Conference on Body Area Networks*.

Jung, S., C. Lauterbach, and W. Weber. 2002. Integrated microelectronics for smart textiles. In *Workshop on Modeling, Analysis, and Middleware Support for Electronic Textiles*, 3–8.

Lehn, D., C. Neely, K. Schoonover, T. Martin, and M. Jones. 2004. e-TAGs: e-Textile Attached Gadgets. In *Proceedings of Communication Networks and Distributed Systems: Modeling and Simulation.*

Lind, E. J., R. Eisler, G. Burghart, S. Jayaraman, R. Rajamanickam, and T. McKee. 1997. A sensate liner for personnel monitoring applications. In *Digest of Papers of the First International Symposium on Wearable Computing*, 98–105.

Marculescu, D., R. Marculescu, N. H. Zamora, P. Stanley-Marbell, P. K. Khosla, S. Park, S. Jayaraman, S. Jung, C. Lauterbach, W. Weber, T. Kirstein, D. Cottet, J. Grzyb, G. Troester, M. Jones, T. Martin, and Z. Nakad. 2003. Electronic textiles: A platform for pervasive computing. In *Proceedings of the IEEE* 91:1995–2018.

Martin, T., M. Jones, J. Edmison, T. Sheikh, and Z. Nakad. 2004. Modeling and simulating electronic textile applications. In *ACM SIGPLAN/SIGBED 2004 Conference on Languages, Compilers, and Tools for Embedded Systems.*

Mattmann, C., O. Amft, H. Harms, G. Troester, and F. Clemens. 2007. Recognizing upper body postures using textile strain sensors. In *Proceedings of the Eleventh International Symposium on Wearable Computers*, 29–36.

Nakad, Z., M. Jones, T. Martin, and R. Shenoy. 2007. Using electronic textiles to implement an acoustic beamforming array: A case study. *Pervasive and Mobile Computing* 3:581–606.

NXP Semiconductors. 2007. UM10204 I^{2}C-bus specification and user manual, rev. 03.

Paradiso, J. A., and T. Starner. 2005. Energy scavenging for mobile and wireless electronics. *IEEE Pervasive Computing* 4 (1): 18–27.

Paradiso, R., G. Loriga, N. Taccini, M. Pacelli, and R. Orselli. 2003. Wearable system for vital signs monitoring. In *International Workshop on New Generation of Wearable Systems for E-Health*, 161–68.

Park, S., C. Gopalsamy, R. Rajamanickam, and S. Jayaraman. 1999. The wearable motherboard: An information infrastructure or sensate liner for medical applications. *Studies in Health Technology and Informatics, IOS Press* 62:252–58.

Sheikh, T. 2003. Modeling of power consumption and fault tolerance for electronic textiles. Master's thesis, Bradley Department of Electrical and Computing Engineering, Virginia Tech.

Shenoy, R. 2003. Design of e-textiles for acoustic applications. Master's thesis, Bradley Department of Electrical and Computing Engineering, Virginia Tech.

Starner, T. 2001a. The challenges of wearable computer: Part 1. *IEEE Micro* 21 (4): 44–52.

Starner, T. 2001b. The challenges of wearable computer: Part 2. *IEEE Micro* 21 (4): 54–67.

Zeh, C. 2006. A flexible design framework for electronic textile systems. Master's thesis, Bradley Department of Electrical and Computing Engineering, Virginia Tech.

7 Humanistic Needs as Seeds in Smart Clothing

Sébastien Duval, Christian Hoareau, and Hiromichi Hashizume

CONTENTS

7.1 INTRODUCTION

Smart clothes, garments empowered by information technologies, may be seen as soft ways to improve the human condition without possessing or burdening people, unlike implants and handheld devices. Although smart clothes are frequently seen from the broader perspectives of wearable* or ubiquitous computing,† their creation requires specific concepts, techniques, and ingredients notably related to textile, electronics, and software. Their proper design and large-scale adoption in various countries in a short time frame also requires the selection of meaningful services and the invention of features appropriate to all members of the general public as initial disappointments may cause massive rejection, incidentally atrophying research and development structures. However, most specialists worldwide still focus on technologies and neglect human aspects, notably due to their excessive focus on the dominant vision of ubiquity (Weiser 1991), which maintains a paucity of knowledge for the exploitation of emerging technologies and for the establishment of high-impact research and development. As a result, the general public may not significantly benefit from existing prototypes, which support few true needs and were mainly thought to be for—sometimes geeky—young male adults, without considering the diversity of potential users and of their environments. Thus, we believe that a different impulse is needed, and that a humanistic perspective will immediately benefit world populations by suggesting new visions and by promoting the creation of more useful and more appropriate smart clothes.

In this chapter, we investigate humanistic aspects in smart clothing globally because we do not know of any such work at the moment. We wish to consider *needs* (aka basic, essential, natural, true needs) and not *wants* (aka artificial, pressing needs) to provide a clear and realistic scope of research: needs are limited, whereas wants are unlimited, and needs are not based on choice although their satisfaction allows choice (Rivers 2008). As a starting point, we consider the works of the founder of humanistic psychology: Abraham Maslow, who is well known for his theory of motivation (Maslow 1987) and its resulting hierarchy of needs, which matches perfectly the aspect we wish to consider. To simplify the discussion, we will, however, limit ourselves here to individuals, and plan to consider groups in depth ulteriorly.

---

* Wearable computing is a field of computer science that concerns computers worn by users, possibly embedded in garments, placed on the body as an accessory or held in hands like cellular phones.
† Ubiquitous computing is a field of computer science that concerns the access and use of computing resources—ideally—in any place and at any time. Wearable computing is seen as a means or part of ubiquitous computing, but ubiquity may also be realized with so-called intelligent environments and, in a more distant future, with implants.

True human needs should constitute the heart of smart clothing and affiliated domains because technologies are meant to serve people wherever and whenever available, permeating physical, social, and digital worlds. Abraham Maslow (1908–1970), the American psychologist at the origin of the humanistic movement, established the existence of a hierarchy of needs that motivates human behaviors (Maslow 1987) by focusing on healthy people rather than on mad ones. Listed by order of decreasing potency, these *humanistic needs* are physiological, safety, belonging, esteem, and self-actualization needs (Figure 7.1); the hierarchy was later extended with a controversial higher level for spirituality. Maslow hinted at the universality of his hierarchy, and his successors strengthened his position, highlighting that the *essence* of the needs seems universal, although their *expression* varies due to, e.g., one's personal history, culture, and environment. Finally, he added that some behaviors are unmotivated, corresponding to the expressions of one's personality and past rather than to a need; for example, one's gait can result from a broken leg that healed incorrectly. Thus, Maslow's theories provide a clear and strong ground to identify and prioritize services and designs worldwide from a human point of view, though with a focus on usefulness that basically ignores art and entertainment.

Physiological needs target homeostasis, such as the need for air to breathe, for a temperature adapted to human physiology, for sufficient high-quality sleep, and for proper nutrition; these needs are the most important because they continuously and permanently relate to immediate survival. *Safety needs* target physical and mental security: the absence of illness or war, freedom, stability, and predictability in one's environment; these needs are also extremely important as they relate to medium- and long-term survival and fitness. *Belonging needs* deal with emotional relationships of all kinds, covering qualitative and quantitative affective bonds with one's family, coworkers, fellows from a sports club, etc. *Esteem needs* have an external and internal component: respect and self-esteem; respect relates to one's image, whereas self-esteem relates to one's perceived and actual abilities and achievements. *Self-actualization* needs were the highest needs in the initial hierarchy of needs (before the addition of a layer for spirituality); they incite us to fulfill our potential, such as becoming a musician. *Self-actualization* needs are extremely difficult to reach because all lower needs have to be satisfied first, because the paucity of attainers

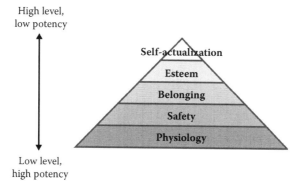

FIGURE 7.1 Maslow's hierarchy of needs.

limits the number of examples and mentors, and because self-actualization is deeply personal, possibly unique to an individual. It is unclear how much technology can help people gratify their self-actualization needs directly.

When considering these five categories of needs, we must remember that lower needs have priority over higher ones; when a need is satisfied, higher needs become salient. However, even partial satisfaction of a need suffices for the emergence of higher needs: priorities evolve gradually. One may discard higher needs when lacking food on a daily basis but, once fed, will start to pay more attention to personal safety. Alternatively, lower needs can come back into focus: when facing a crisis such as a job loss or divorce, one can drop to the lower level that reflects what was lost. After losing a job, for example, one may stop meeting friends and focus instead on job-hunting to secure food, housing, and medical treatments; after a divorce, one may stop focusing on personal development and instead resume former relationships or meet family members. Salient needs impact our current perception of the environment and future, which can easily lead to a crisis due to the underestimation of lower, currently satisfied, needs.

Scholars, inventors, and artists cooperate in many countries to provide humanistic products, services, and environments, as exemplified by modern sportswear, several airports, and several libraries. However, such dynamics usually lack when electronics and software are embedded, probably because many competent specialists prefer traditional ways, and because many technologists assume that non-technological issues are trivial, outside their scope, or easy to incorporate at the final stage. The neglect of human factors pervades computing-oriented visions of the future and ubiquity-related prototypes: they often feature artificial or even trivial needs (e.g., adapting music to one's walk, or making coffee on demand) and target Western lifestyles (e.g., RFID* tags scattered in a *big* house, navigation data for *cars*, recommendation systems based on *individual* rather than collective wishes). This clear trend for handhelds such as cellular phones foreshadows likely—unsatisfactory— conceptions of smart clothes.

In 1991, Mark Weiser defined and vulgarized a technology-centered vision of ubiquity that spread worldwide and became the most popular among specialists (Weiser 1991) because it incorporated leading-edge technologies without being disruptive, and gave the community clear and extensible objectives that seemed realistic and attainable on the short-run. Unfortunately, it focused on labor rather than quality of life, on quantitative rather than qualitative benefits, and on sexless, supposedly acculturate, healthy young adults. If progress had quickly occurred, stakeholders would probably have considered other issues, and the partiality of this vision would have been inconsequential. As things happened, though, scientists got stuck for almost 20 years with the scarcity of wearable energy and with the complexity of context awareness, in a fashion reminiscent of the demise of artificial intelligence. In 2008, the community still failed to acknowledge the situation, reallocate its resources, and consider alternative visions of the future that would at least benefit the general public in the short term. This situation may persist for years because established scientific

---

* RFID, radio frequency identifier; RFID tags are typically used to wirelessly identify nearby objects or to locate an object in a building.

conferences and journals maintain a monolithic agenda marked by a search for efficiency and productivity, and by the predominance of technology.

However, a few researchers challenge the current vision and champion an alternative future, while practical advances in smart clothing provide a basis to rethink applications. For example, Genevieve Bell (Bell, Blythe, and Sengers 2005) and Paul Dourish (Bell and Dourish 2007) highlighted that the world is moving fast in a direction different from that pointed out by Mark Weiser, with many deployed services outside the topics usually considered by the community, such as religious services. Besides, researchers in smart clothing are discovering that the applications of smart clothing should differ from those of handhelds: Who wants garments to support navigation when cellular phones already do it? Smart clothes should rather provide meaningful dedicated services such as physiological monitoring and physical training/rehabilitation. Hopefully, human-centered approaches will be readily adopted in smart clothing because non-technologists such as textiles specialists and fashion designers already participate. These approaches should notably cover neglected human issues such as everyday comfort and functionality, personal and social image, dignity, efficacy, dependence, and human growth.

In the following sections, we discuss the absence of a comprehensive and coherent vision for smart clothing, first criticizing Mark Weiser's technology-centered vision, then indicating current dynamics for mobility and ubiquity worldwide, and finally showing crucial problems in governmental initiatives. Then, we evaluate the nature of humanistic clothing that may be developed shortly, presenting promising prototypes for physiological needs as well as unexploited concepts for safety needs, and the lack of motivating concepts for belonging and esteem needs. Exploring the reactions of the general public towards smart clothes in two different cultures, we show that wishes and concerns are predictable from, or compatible with, Abraham Maslow's hierarchy of needs, and briefly discuss demographic variations. To show the importance of human diversity, we detail the case of children and elders. Based on these investigations and analyses, we present our humanistic vision for smart clothing, and indicate factors that may be critical to its realization. Finally, we conclude on perspectives related to theory and practice.

7.2 NOBODY HAS ESTABLISHED A COMPREHENSIVE AND COHERENT VISION FOR SMART CLOTHING YET

So far, researchers have investigated smart clothing without guidance from a dedicated vision that could organize and unite research and development for the general public locally or globally, notably due to greater attention given to the related fields of *wearable computing* and *ubiquitous computing*. Dominated from the start by technologists, these fields are strongly influenced by Mark Weiser's vision of the future based on technologies already available 20 years ago (Weiser 1991) but also on unresolved algorithmic issues; the absence of successive or competing visions limited these fields to *technical challenges for better productivity*. Although the scientific community remained stuck on these challenges, the world moved on, and original meaningful services appeared in several countries thanks to entrepreneurial creativity, to cellular phones, to the Internet, and to great infrastructures in densely populated areas such as Singapore and South Korea. Aside from this, most governments focused on infrastructures for the emergence of services, and typically on military and medical applications, with notable efforts towards care for elders due to population aging. These well-funded governmental efforts were—and still are—meaningful for security, societal, and moral reasons but failed to provide fundamental bases for civil smart clothes. Besides, many funds were wasted by intergovernmental competition. Consequently, specialists in smart clothing worked on well-funded or personally interesting topics, adopting Mark Weiser's vision of ubiquity as an acceptable loose guide.

7.2.1 Mark Weiser's Partial Vision Froze without Adaptations for Smart Clothing

International research in ubiquitous computing is still significantly guided by the dated vision of Mark Weiser (Weiser 1991; Weiser and Brown 1997), which was only recently challenged by Genevieve Bell and Paul Dourish as incomplete and as increasingly disconnected from the reality of systems used and requested worldwide (Bell and Dourish 2007). Mark Weiser centered his vision on *productivity*, *efficiency*, and *labor*, whereas systems adopted worldwide now mainly concern latent potentials, the quality of experiences, remote communication, and everyday life. He focused on technology, with privacy-related caveats, creating a twisted vision that obliterates humanity; although this has been recognized, no vision emerged to change the situation. Mark Weiser's vision evokes machines that know when to prepare coffee, environments that monitor children, and artifacts that transmit data when waved towards a recipient or device; although smart clothes are not evoked, the scenario leaves place to the imagination and readers frequently visualize many more situations, services, and devices, including body implants, digital jewelry, and smart clothes for both everyday life and work.

In his seminal paper (Weiser 1991), Mark Weiser illustrated his ideas with a scenario that involved "Sal," a woman living an American way of life, with a house, a car, children, and an informal work environment. In this scenario, ubiquitous technologies provide novel services dealing with context awareness, location, and productivity, but the author also insisted on technologies directly improving lifestyles, notably by preparing coffee (!) at the right moment, an example evoked twice in his one-page scenario. The context-aware services are activity indicators, biographical repertories, events archives, and weather indicators. The activity indicators notably inform Sal about what happened in her neighborhood when she wakes up, and tell her when a meeting occurs at her office. The biographical repertories furnish information about a person briefly met during a meeting. The events archives indicate a colleague's request for discussion, the content of a document, and the list of attendees at a meeting. Finally, the weather indicators displayed on Sal's office windows indicate the current weather, provide forecasts, and reveal the weather at distant offices. Location services monitor children at home, follow colleagues at the office, and disclose the location of a lost garage door opener manual. Productivity services prepare coffee at home and at the office, transmit information such as news quotes and reminders, and support transportation by indicating traffic jams and available parking spaces. This should convince readers that the vision focuses on productivity, efficiency, and labor. However, in a complementary paper (Weiser and Brown 1997), the scope of visible applications is extended with additional services for corporate finance, school homework, and commerce, e.g., knowing whether a t-shirt is still on sale or not, an example that apparently did not seem ludicrous to the research community.

The services proposed in these two publications barely relate to humanistic needs: none of the examples relates to physiological needs, a single service may relate to safety needs (locating children), and only one service may relate to belonging needs (displaying activities in the neighborhood as it could foster a community sense). However, Mark Weiser apparently does not cite these services to highlight a potential to satisfy safety and belonging needs because the children's location is only used inside, and not, e.g., near a road where children could have an accident, and neighborhood information is just consulted, without any hint that Sal would exploit it to contact or help her neighbors. Similarly, locating objects could be vital (e.g., drugs) but finding the manual for a garage door is trivial and unlikely to serve twice unless users form bad habits due to the reliability of ubiquitous technologies. Thus, the scenario neither deals with, nor hints at its ulterior extension to, humanistic needs.

After proposing his initial vision (Weiser 1991), Mark Weiser proposed calm computing (Weiser and Brown 1997) as a complement, arguing that it would respond to what he identified as two human needs: *relaxation* and *information*. Although relaxation relates to human physiological and safety needs, he fails to include this in his vision because he discusses *systems that do not unnerve* rather than *systems that calm users or bystanders*; he therefore does not propose to improve situations when introducing ubiquity but rather to avoid worsening them, a weak proposal. As for information, it seems beside the point because it is more a tool than a—true—need. Finally, Mark Weiser highlighted privacy issues, which relate to safety, but discussed it only as a consequence of ubiquity instead of a problem that ubiquity could solve, once more cautioning against technologies rather than considering their potential benefits. As a result, this vision falls short of any humanistic dimension.

Unfortunately, Mark Weiser passed away in 1999 without his influence waning in the following 10 years: he remains widely cited in conferences and journals, and his scenario is still a starting point for international workshops. Besides, numerous scientific publications cite his vision in their introduction, year after year, with no critical analysis of its current pertinence for the community and for the world at large. Similarly, specialists in smart clothing still have to propose adapted or novel visions to match the specificities of their equipment.

7.2.2 THE WORLD HEADS ITS OWN WAY IN REGARD TO MOBILITY AND UBIQUITY

One can be tempted to say that ubiquitous computing is already here if one looks at the spread of mobile technologies such as cellular phones, at the presence of large-scale displays in capitals such as Tokyo (Japan) and Seoul (South Korea), at the availability of smart cards for transportation in Paris (France), or at visual codes fixed on buildings to guide taxis towards customers in Singapore. The fact that all these elements escaped Mark Weiser's vision suggests that researchers are victims of their ivory towers. As first convincingly demonstrated by Genevieve Bell, the evolution and diffusion of ubiquity worldwide differ much from Mark Weiser's descriptions and thus from the scientific community's current landmarks and goals (Bell and Dourish 2007).

Noteworthy elements include first the state of current technologies, far beyond what Mark Weiser accessed when he produced his vision of "The Computer for the Twenty-First Century": computers are orders of magnitude faster; cellular phones incorporate e-mail, Internet access, cameras, music players, localization based on Global Positioning Systems; and mobile devices are commonly used for transportation or to pay at shops in Japan and South Korea. Besides, cellular phones are spreading even in least advanced countries and among children below 10 years old! Novel devices and infrastructures such as Apple's iPhone® and Google's Android™ may incidentally dynamize the mobile field regarding hardware, software, and services.

Second, numerous successful services are deeply rooted in local environments, cultures, and religions, and involve diverse users. For example, the Singaporean government created online shrines were one could pray for ancestors during the 2003 SARS outbreak; local mobile services also include *feng shui* advice and translation for the four official local languages (Bell and Dourish 2007). In South Korea, most children above 6 years old and adults above 60 years old regularly use Internet or mobile services, a greater diversity of users than what Mark Weiser evoked. Elsewhere, one might also find displays of "prayer times on a mosque wall" (Bell and Dourish 2007) or mobile alert services for tsunamis or earthquakes that would not make sense in, e.g., France.

Third, ubiquity is well implanted in several densely populated areas such as the whole island of Singapore and the capital of South Korea. Accesses to wireless networks are pervasive in Singapore, and services include localization information for taxis to reach customers, smart cards for trains, and camera monitoring for safety in public transports and public places. In South Korea, 80% of the inhabitants above 6 years old access Internet and mobile services, although more frequently from home than on the move (Bell and Dourish 2007). Ubiquity is so present that it disappears from the conscience of users, coming back to mind during trips to the countryside. In many industrialized countries also, personal digital assistants (PDAs) and portable game consoles complement cellular phones, and large-scale displays are visible in the streets, eventually alerting about earthquakes in, e.g., Japan.

Interestingly, critical features of infrastructures and services also rely on values that differ much from the American ones suggested by Mark Weiser's vision. For example, censorship is widespread and apparently well accepted by the population in Singapore, and practices have a collective dimension instead of—or in addition to—an individualistic dimension (Bell and Dourish 2007). Besides, Mark Weiser envisioned clean, seamless digital environments in stark contrast with the messiness of the real world; for example, cellular phones use incompatible protocols that sometimes prevent usage abroad. Similarly, international travelers face the absence of common wireless providers in different countries, requiring changes in configuration, manipulations in foreign languages, and payments to unknown or non-trusted entities. Within countries too, connection accesses vary: traveling to the countryside, in mountains, in subways, and in high-speed trains makes digital experiences far from seamless. In the end, the world is neither beyond nor behind Mark Weiser's vision: it is just differently shaped!

7.2.3 GOVERNMENTS LACK A HOLISTIC VISION TO GUIDE THEIR INVOLVEMENT IN SMART CLOTHING

Ubiquity-related projects have started all over the world during the past 10 years, once Internet had convinced governments of the potential success of emerging information technologies. Some governments focused on military and medical research, e.g., the United States of America, while others focused on future infrastructures to enable and empower smart artifacts and smart environments, e.g., the European nations through European Framework Programs, or on high-speed connections, location-based services, and citizen-oriented services, e.g., Japan and South Korea. Besides, these governments also typically funded research to support the elderly in the context of rapidly graying populations and shortages of qualified medical staff. Thanks to an ambitious governmental plan for the development of technology, ubiquity became a reality in Singapore, and more projects have already started. All these worldwide efforts are encouraging but lack a holistic perspective, with impacts on crucial elements such as sustainability, humanistic needs, and international collaborations.

Although ubiquity could be a boon to the environment, this aspect remains clearly out of the scope of mainstream projects. This is particularly regrettable in the case of smart clothes, which offer much space to embed dedicated sensors: smart clothes could easily monitor surrounding temperatures and ultra-violet radiations, and ideally also detect allergens and pollution, immediately benefiting wearers, especially weaker people such as asthmatics, pregnant women, children, and elders, and possibly people at large if information were shared with adequate agencies or institutions. Raising awareness, such garments could lead to behavioral changes and thus to more responsible lifestyles. In addition, smart clothes should be recyclable and exploit renewable sources of energy, but research on these aspects is insufficient and unlikely to progress without governmental incentives or impulses.

Human aspects are also neglected as research on ubiquitous technologies satisfying humanistic needs remains rare, with the notable exception of medical and safety support for the elderly. Smart clothes, however, seem extremely promising because they can contain sensors to monitor health and movements, and actuators to inform and provide emergency responses, because they are difficult to forget, contrary to accessories or handheld devices. Initiatives for higher needs are cruelly lacking. Humanistic needs should be considered at the governmental level because some important services require enormous investments, because some users and services may be neglected otherwise, and because infrastructures should be designed and built with humanistic needs in mind.

Finally, governmental programs tend to leave aside the highly interdisciplinary and potentially disruptive nature of smart clothes. The required collaborations between specialists in clothing, computer science, electronics, fashion, materials, medicine, sociology, and psychology are unlikely to arise spontaneously due to the number of domains involved, due to the usual cultural clashes between specialists from different fields, and due to the costs involved in bootstrapping the research. Consequently, research centers for smart clothing are rare, the level of achievements is frequently disappointing to external observers, and human issues are typically neglected.

7.3   SMART CLOTHES AT HAND FOR THE WORLD PUBLIC GRATIFY FEW HUMANISTIC NEEDS

Specialists of different fields may envision smart clothes with divergent focuses and priorities, for example, everyday comfort and functionality in clothing, personal and social image in fashion, dignity and efficacy in medicine, growth and dependence in psychology, reliability and security in electronics, or awareness and empowerment in wearable computing. Researchers investigated concepts of smart clothes without integrating these aspects, neglecting a host of human issues critical for the general public in everyday life, and strongly biasing realizations towards geeks, patients, and soldiers, e.g., (Zieniewicz et al. 2002). Hereafter, we examine core concepts of smart clothes based on Maslow's psychological theories, focusing on individuals rather than groups, and show that some important concepts lack strong bases, most others requiring significant investigations for a successful adoption worldwide. We do not intend to exhaustively review smart clothes from a humanistic perspective, and may have missed a few innovative works, but we intend to show where the community stands and where it may go or not depending on its collective choices.

7.3.1   SMART CLOTHES GRATIFYING PHYSIOLOGICAL NEEDS ARE THE MOST PROMISING IN THE SHORT TERM

Maslow identified physiological needs as the most important set of human needs as they continuously and permanently concern immediate survival, and as they relate to health on a longer time frame. Unsurprisingly, smart clothing and wearable computing specialists have been researching and developing dedicated prototypes in several countries during the past 20 years as technological components, perceived markets, governmental funds, and moral imperatives clearly converged. Interesting smart clothes were proposed to save lives and support physical fitness, as well as wearable devices to improve nutrition and maintain homeostasis, but systems related to sleep and air quality are remarkably absent.

Smart clothes monitoring wearers' cardiac and bodily activities are probably the most publicized, and may save the lives of patients suffering from physical conditions by advising them about impending problems, alerting emergency crews after dramatic events, and informing doctors or families about long-term trends. Sensatex designed the SmartShirt (Sensatex 2005) to monitor patients' health with various sensors, but systems may also be used by, e.g., sportsmen. Vivometrics apparently had different customers in mind as its LifeShirt (Vivometrics 2005) was proposed to continuously monitor the physiological states and postures of firefighters, functions that could be greatly combined with sensors assessing environmental temperature and toxicity to inform on-site colleagues and command centers (Jiang et al. 2004). Beyond monitoring and informing, medical clothes may even help patients maintain or resume an acceptable cardiac activity by automatically injecting drugs (Jafari et al. 2005) based on real-time cardiac data matched to personalized profiles. These systems are not yet well integrated into textiles but this is negligible considering the potential benefits for the aged, obese, or those who live unhealthy lifestyles: they may prefer a reduced clothing comfort to death. However, providers may have to adapt medical vests to ensure sufficient comfort and performances in varied climates and environments. Besides, researchers have also suggested smart clothes for pregnant women (Bougia, Karvounis, and Fotiadis 2007), potentially sharing monitored fetal activity with the father and doctor, pajamas designed to prevent the sudden infant death syndrome (Verhaert 2007), and shoes monitoring the health of premature babies.

Besides, smart clothes may continuously favor healthy lifestyles of diverse people worldwide, starting with exercise and nutrition (Amft, Hunker, and Troster 2005; Kawahara et al. 2005), with a few simple and cheap sensors. High-grade cardiac sensors being overly expensive for non-medical services, engineers may here exploit less reliable equipment to reduce costs. Accelerometers or stretchable conductive textiles could monitor the amount and diversity of movements at joints; thermometers in contact with the skin or the air could monitor the temperature of the body and environment; and light sensors could monitor ultraviolet exposure at meaningful locations depending on the clothes worn. Such garments would require little energy, especially if some is harvested during exercise, so the only real difficulty is to sense body temperature: skin temperature differs from inner-body temperature and changes at a different speed. Ideally, lifestyle-enhancing garments would also monitor food intakes, maybe with RFID readers or cameras mounted on the chest or sleeves, extending mobile services such as Lifewatcher (Lifewatcher 2008). However, the task is not trivial, and results would highly depend on food distributors (e.g., placing RFID tags in all packages), restaurateurs (e.g., providing details for all meals), and users themselves (e.g., cheating their own support system). Wearers could then receive advice based on generic recommendations by health organizations or personalized recommendations by companies, possibly integrating cultural or religious specificities. Depending on situations, smart clothes may also alert doctors or family members and, if acceptable to wearers, automatically tailor orders in restaurants or request food delivery for home.

Some smart clothes may also support homeostasis, but applications are globally limited by the ignorance of inner-body information and unsatisfactory algorithms. Smart clothes may, for example, successfully cancel heat stress by cooling down (United States Army 2005) or warming up (Rantanen et al. 2002) the body based only on measurements of the environmental temperature. However, some applications currently appear not feasible because they require blood tests or cumbersome equipment. Occasionally, elegant solutions are found, like the GlucoWatch (Tierney et al. 2001), a watch that monitors diabetics' glucose levels non-invasively; in some cases feasibility, reliability, or convenience may instead lead to the use of implants communicating with smart clothes or mobile devices.

If we now turn to the least physically active period of our life, we may consider the concept of smart pajamas. Smart pajamas may offer services similar to daytime smart clothes, such as monitoring cardiac activities and maintaining body temperature, but may also provide unique services, for example, soothing children or monitoring sleep. Monitoring sleep with smart clothes is meaningful because we usually ignore what happens to us while we sleep, because smart clothes may uncover sleep troubles, and because it may save lives. One scary sleep-related trouble is *sleep apnea*, a repeated cessation of respiration during sleep; this disease can result in strokes and death (Dement 1976) but affected people frequently ignore their situation. A watch has already been proposed to monitor sleep (Sleeptracker 2005) but garments allow much more: detecting respiration patterns, body movements, and snores that characterize specific sleep troubles. In addition, sleep specialists could exploit daytime information, for example, about exposure to light, which affects the biological clock and thus sleep. Smart pajamas may soon be designed with stretch sensors and a single microphone, but dedicated algorithms will be required to detect potential troubles.

Finally, smart clothes could valuably monitor air quality, informing wearers, bystanders, and the world about the presence of allergens and pollution at natural sites, in streets, and inside buildings. Researchers have already tried to measure and visualize exposure to pollutants with handhelds (Kaur et al. 2006) and detectors mounted on bicycles, but sensors are too big to fit in everyday garments; unfortunately, miniaturization seems difficult for physical reasons. Anyway, smart clothes may still inform wearers and bystanders and thus change the world: networked garments may alter their color or shape according to local pollution indicators, and context-aware garments may similarly react depending on the estimated carbon footprint of their wearer, possibly leading to behavioral changes and environmental improvements.

7.3.2 Smart Clothes May Gratify Safety Needs after Intense Research and Development

Maslow identified safety needs as the second most important set of human needs as they relate to survival in the medium term and long term. Unsurprisingly, mobile and wearable computing specialists have researched and developed diverse dedicated systems, some of them already available with standard cellular phones (e.g., 3D guidance after earthquakes in Japan). Surprisingly, however, specialists have proposed almost no safety garment significantly more valuable than a simple cellular phone or dedicated wearable device. Some prototypes may be attributed to physiological or safety needs based on their expected probability of use: cardiac monitoring would belong to physiological needs for patients but "only" to safety needs for healthy joggers; we present below smart clothes reasonably dedicated to safety, considering stress, then criminality, and finally accidents and natural dangers.

Smart clothes may relieve stress, improving quality of life and helping fight against depression, by massaging wearers, diffusing relaxing smells, or producing soothing music. Providing massaging garments is not trivial because mechanical components are cumbersome, which poses significant problems of integration with other functions and of adaptation for children and elders. Similarly, diffusing relaxing smells is a technical challenge as odors should reach the wearer rather than bystanders, in specific quantities for specific durations, and containers for chemicals should be easy to refill and resist shocks; such research has started (Smart Second Skin 2008) but everyday uses are still out of reach. On the contrary, everyday garments and pajamas may easily soothe wearers with music adapted to circumstances, using physiological and movement sensors.

Criminality is a less joyful issue but may also be partially dealt with using dedicated smart clothes. Usable in most places, the StartleCam (Healey and Picard 1998) is a remarkable safety-related wearable-computing concept: the device sends photos of scenes facing the wearer to trusted people or emergency services when it detects physiological reactions attributed to fear. The prototype seems unreliable because it neglects external factors influencing physiology, but smart clothes may perform well thanks to sensors evaluating, e.g., physical activities, temperature, humidity, or the presence of lovers; tests would clarify the practical performances and usefulness of such garments. Besides, smart clothes could incapacitate offenders, e.g., by providing electric shocks, or mark them with harmless but persistent chemicals to ease identification by policemen; simple devices already do this but may be unreachable when needed, contrary to worn clothes. The adoption of aggressive garments would critically depend on reliability, personality, culture, and laws.

Less controversial, garments dedicated to accidents or natural disasters would be more readily adopted. For example, bikers and elders may remain in better health following falls thanks to garments inflating locally or completely to absorb shocks, people may avoid drowning after tsunamis or falls in rivers thanks to garments with reactive floaters, and skiers may be found more quickly after avalanches thanks to color-changing or illuminating garments. These niche prototypes may be rapidly developed and commercialized due to the concerns of specific groups worldwide, funding improvements. Targeting broader populations may paradoxically fail due to lower risk perception; for example, garments indicating escape routes with whole-body tactile stimulations would usefully complement services on cellular phones because they would work even when smoke is thick and alarms loud, but who seriously thinks about risks of fire even once a year? Multipurpose garments are more attractive but require much research, either due to advanced functions or to difficult integration. An example of advanced function is the automatic detection of hemorrhages and twisted or broken limbs of unconscious victims of a disaster, for transmission to rescuers along with location data. Integration is typically difficult when functions rely on different technologies; for example, physiological sensors evaluating health may compete for space and weight with tactile actuators guiding wearers during fires. These concepts remain globally unexplored, although they are unaffected by a major limitation of smart clothing: they require little energy because functions are not used continuously.

7.3.3 Novel Concepts and Fundamental Psychological Bases Are Required for Higher Needs

Maslow identified belonging, self-esteem, and self-actualization needs as the three highest sets of needs, and highlighted that they concern relatively few people because lower needs must be gratified for higher needs to emerge. Existing networked devices greatly support belonging by establishing and maintaining distant bonds with, e.g., text/voice/video direct communication, status/diary dissemination, etc. However, mobile devices do not gratify many self-esteem and self-actualization needs, notably because most people wrestle with lower needs, and because even psychologists ignore how to gratify self-actualization needs. Smart clothes may complement hand-held devices by providing emotional or tactile information, naturally exchanging personal information, and establishing community bonds. However, more ideas are required to gratify the general public's belonging and self-esteem needs worldwide.

Exploiting physiological sensors, smart clothes may inform about emotions during face-to-face or phone discussions, or even on demand as a background service for acquaintances. Sensor-based gloves disclosing arousal locally (Picard and Scheirer 2001) found few owners due to their low reliability and to psycho-social worries, but garments may succeed as they can better react to the context and better evaluate emotions using sensors for, e.g., physical activities, temperature, and humidity. Whole-body stimulations could similarly extend communication, for example, allowing mothers to hug their children at a distance. Besides, garments may automatically exchange data to favor long-term contacts (Kanis et al. 2005), using movement sensors to detect when people shake hands or bow. However, cultural mores may doom such services in some places, for example, because exchanging business cards has strong meanings in Japan. Finally, garments could establish community bonds, as formerly attempted with badges (Borovoy et al. 1998; Falk and Bjork 1999); in addition to displaying community-relevant information, garments could change their shape or color to represent a facet of the wearer adapted to bystanders, or even produce dedicated smells or sounds depending on the community, location, and event.

7.4 REACTIONS OF THE FRENCH AND JAPANESE PUBLIC SUGGEST HUMANISTIC NEEDS ARE GOOD SEEDS

In 2005 and 2006, we gathered surface information about the general public's perception of potential everyday life uses and features of smart clothes in France and Japan, evaluating the influence of humanistic needs with informal interviews, self-completion questionnaires, and use-cases. Three hundred twenty-nine males and females aged 14 to 67 years old, mainly French and Japanese, participated in this pilot study. Featuring leading-edge technologies easily accessible to our group, Japan was appropriate to evaluate people's interest in smart clothing and collect sufficient quality information to prepare future large-scale investigations. Nurturing a vastly different culture and homeland of our main investigator, France was selected as a suitable complement to test claims of universality for our hypotheses.

Background information of interest included Internet use and the possession of mobile systems such as cellular phones and car navigation support, which revealed that our respondents usually used the Internet daily and possessed cellular phones as well as laptops. Core issues were the acceptance of services gratifying fundamental needs, differences in reactions for services gratifying fundamental and artificial needs, wishes for critical features, and the validity of results worldwide.

We established our first hypotheses based on the seminal works of Maslow (Maslow 1987), explored them and identified critical features with interviews, tested our hypotheses with questionnaires, and clarified findings with use-cases. To realistically manage our resources, we exploited non-random samples of the French and Japanese populations, and restricted our investigations to lower needs (physiology, safety, and belonging), which are by definition the most important, and for which we can already develop credible solutions. The reactions of respondents expressed a clear pattern compatible with Maslow's theories, presented after the details of our methods along with demographic variations.

7.4.1 We Exploited Three Complementary Methods to Favor Meaningful and Valid Results

Because little data is available about the general public's perception of smart clothes, both broad and deep investigations are desirable worldwide in the short term, e.g., based on anthropological studies (observation in situation), thought experiments (scenario, interview, questionnaire), or use-cases (evaluation of prototypes). Although useful to study diverse contexts and the introduction of technologies, anthropology is too limited to guide the creation of disruptive technologies. Researchers can extract more value from thought experiments as existing artifacts are unnecessary, most issues can be raised, most people can participate, and data can be quickly gathered in several locations. However, thought experiments provide less reliable data and easily miss critical elements because they lack factual roots and are channeled by researchers. Finally, use-cases can easily provide deep and subconscious information, although narrowed by the costs of prototyping. Thus, we decided to investigate general issues in big groups with interviews for qualitative data and questionnaires for quantitative data, and specific issues in small groups with use-cases.

First, we informally interviewed 26 psychologists and computer scientists for approximately 30 minutes in France and Japan to discuss their views on humanistic needs in smart clothing, and to identify their main concerns regarding adoption by the general public. We consequently realized the importance of human diversity and environments but assigned these factors to ulterior studies requiring additional resources. Practically, we noted deep concerns regarding the monitoring of wearers with sensors, and the autonomous control of functions by smart systems.

Second, we prepared a questionnaire dedicated to the general public evoking lower fundamental needs, artificial needs, monitoring, autonomy, and features typically suggested by the *wearable computing* community such as the recognition of a wearer's emotions; because respondents would be unfamiliar with smart clothing, we focused on easy-to-imagine enhanced garments in everyday life. French and Japanese native speakers checked the questionnaire, and we rephrased items, removed technical terms, and added an introduction according to the comments of a

TABLE 7.1
French and Japanese Respondents

	French Men	French Women	Japanese Men	Japanese Women
Number of people	115	59	61	54
Age range (years)	14 to 67	14 to 58	19 to 54	14 to 45
Age mean (years)	≈26	≈25	≈29	≈30

pilot group, which led to two pages including seven series of closed-ended items and an open-ended question. For closed-ended questions, the participants rated assertions as 1–strongly disagree, 2–disagree, 3–neither agree nor disagree, 4–agree, and 5–strongly agree. We considered a mean below 2.5 significant for rejection and above 3.5 significant for acceptance. Details about the 289 respondents are available in Table 7.1.

We mainly distributed paper questionnaires in 2005 in train stations, bars, and coffee shops on weekdays and weekends, and digital ones in universities and via a public relations office, which provided large samples with moderate randomness. Respondents included artists, designers, librarians, reporters, students, teachers, researchers, engineers, secretaries, salespeople, managers, housewives, retirees, medical staff, soldiers, and preachers. We did not show them any photo or video of systems to avoid bias, but introduced the study as *research on new technologies: clothes possessing particular features, capacities, and some kind of intelligence*, and indicated that prototypes were being designed in France, Japan, and the United States of America. In public places, respondents usually answered in 15 minutes.

Third, we clarified results about belonging needs with use-case experiments in Tokyo (Japan) in 2006 with 14 Algerian, French, German, and Japanese citizens aged 21 to 32 years, including two females, who simulated encounters at professional seminars for 10 minutes while wearing an enhanced jacket (Figure 7.2). Passersby could see a back screen embedded to attract them and a front screen to support conversations with personal information of possible mutual interest, but the jacket also contained a touch-sensitive interface and sensors for heartbeats and skin conductivity to visually reflect emotions. Made solely for our experiments, this jacket did not need to be appealing or look like a final product.

7.4.2 The General Public Requests Clothes in a Pattern Matching Maslow's Theories

Looking at the acceptance of services gratifying fundamental needs, the French and Japanese general publics appear strongly attracted by smart clothes that satisfy the lowest human needs: services for physiological and safety needs are on the average rated 3.5 to 4.5 out of 5 points (Figure 7.3). Not surprisingly, the respondents rate emergency services highest but they also highly appraise garments that maintain body temperature, monitor sportive activities, or analyze the air, which reflects

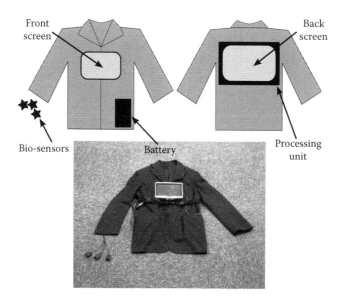

FIGURE 7.2 Wearable system created for the experiments.

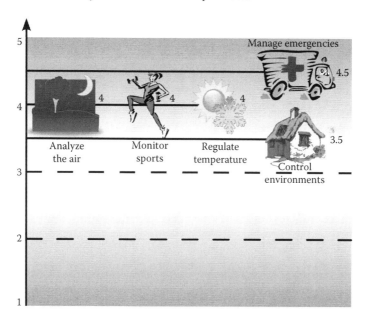

FIGURE 7.3 Acceptance of services for physiological and safety needs.

both unfulfilled needs and the acceptability of technological solutions according to respondents' comments; the study of Edmison et al. (2005) on the perception of medical wearables complements and validates further these results.

Examples of assertions rated by respondents: "It would be acceptable for me to wear clothes that analyze the air (smells, pollution, temperature)," and "I would

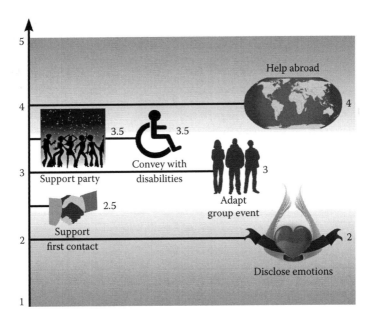

FIGURE 7.4   Acceptance of services for belonging needs.

agree to use garments that monitor my condition (heart beats, movements) to adapt my environment to my needs (temperature, light, music)."

The respondents are less attracted by smart clothes that satisfy the next level of human needs: services for belonging needs are rated between 2 and 4 out of 5 points (Figure 7.4). We predicted lower acceptance of smart clothes for belonging needs than for physiological or safety needs because Maslow classified needs by order of prepotency, and thus related services by order of potential desirability. The respondents predictably showed interest in support for disrupted communication, for example, due to linguistic problems during trips (score: 4) or physical issues with disabled people (score: 3.5). However, the rejection of support for first encounters (score: 2.5) and of emotional disclosure (score: 2) required clarifications; we therefore designed our enhanced jacket (Figure 7.2) and used it to investigate these two discrepancies.

Examples of assertions rated by respondents: "Enhanced clothes would be useful to communicate with disabled people," and "I would agree to use garments that monitor my condition (heart beats, movements) to reveal my emotions to surrounding people."

Fourteen participants filled out questionnaires before and after interacting together while wearing the jacket, which revealed common interests by displaying graphics based on personal information and revealed arousal (i.e., excitement or anxiety) by modifying the background color based on their heart rate and skin conductivity. The usual rejection of support for first encounters was explained by a lack of experience with actual systems: perception rose on average by 1 point out of 5 during the experiments. The rejection of emotional disclosure, however, proved to be mainly due to a perceived danger and a perceived uselessness (Figure 7.5). As the perception of danger is tantamount to actual danger, we concluded that emotional disclosure is rejected

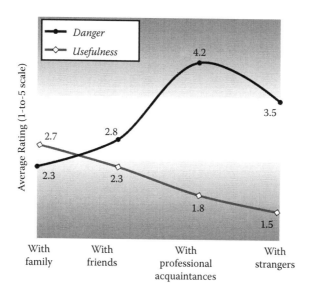

FIGURE 7.5 Perceived danger and usefulness of emotional displays.

due to the non-respect of safety needs, which matches and confirms Maslow's stated priority of lower needs over higher needs (safety over belonging in this case).

As indicated previously, we did not investigate higher fundamental needs (i.e., self-esteem and self-actualization) due to our limited resources. However, we compared the reactions for services gratifying lower fundamental needs to those for artificial needs. Overall, smart clothes fulfilling artificial needs were considered neutrally, neither requested nor rejected. Thus, the respondents globally requested smart clothes gratifying fundamental needs, ignored smart clothes gratifying artificial needs, and rejected smart clothes clashing with fundamental needs. This does not mean that smart clothes for artificial needs will not be adopted; it just means that their adoption will be more contingent on individual interests and contexts, and on costs.

To complement these results about services, we considered wishes for critical features: monitoring, autonomy, and emotional disclosure. Unexpectedly, the respondents significantly accepted physiological monitoring, although our interviewed specialists feared that privacy risks would lead to its rejection; however, a negative shift in perception may still occur due to inappropriate commercial practices or insufficient security after the dissemination of related smart clothes. Regarding autonomy, the respondents globally rejected full control of the functions by an artificial agent but the preferred solution seemingly depends on culture: the French requested full user control whereas the Japanese requested limited artificial intelligence (Figure 7.6). Open comments suggest that respondents rejected full control due to fears of physical and social harm resulting from the action of artificial agents, for example, a bad control of temperature leading to sickness or the display of embarrassing photos to one's acquaintances. Wishes for autonomy are thus highly influenced by safety needs. Finally, respondents rejected emotional disclosure; the intensity of reactions varied with the social distance to those who may learn the emotional states (Figure 7.5),

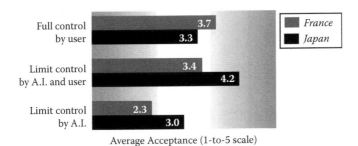

Average Acceptance (1-to-5 scale)

FIGURE 7.6   Acceptance of A.I. control based on perceived danger.

but other factors participate: local disclosure, e.g., during face-to-face discussions is more feared than remote disclosure, e.g., on the phone.

Examples of assertions rated by respondents: "If I had enhanced garments, I would like them to be controlled by some form of artificial intelligence," and "I would agree to use garments that monitor my condition (heart beats, movements) to reveal my emotions to surrounding people."

All these results are either predicted by, or compatible with, Maslow's theories, and suggest services to develop in priority, at least in France and Japan. However, individuals should more or less desire specific smart clothes depending on their current level of satisfaction of fundamental needs, which varies with one's condition, environment, and history, and on their interpretation of the appropriateness of proposed services, which varies with actual experiences and with culture. We notably expect corporate and governmental policies to shape long-term adoption as privacy issues are well known for cellular phones and online medical services; a deep concern in this respect is the absence of strategies for designers and researchers to manage diverse users, contexts, and legal frameworks worldwide. Hereafter, we briefly discuss demographic variations, notably due to culture.

7.4.3   Demographic Groups Assert Non-Uniformity at the Edges of the Pattern

To test the universality of our results, we collected data from males and females in two different cultures, leaving age to ulterior studies because involving children and elders would have required many more resources. Maslow suggested that the *essence* of needs is universal, whereas their *expression* is variable; we therefore correctly expected males and females in France and Japan to perceive smart clothes gratifying fundamental needs similarly, albeit with differing intensities. Before discussing demographic variations, we would like to highlight that results attributed to culture may instead be due to geography, history, or resources: Japan is harsher than France due to earthquakes, typhoons, and tsunamis; behavior in Japan is influenced by laws promoted by the winners of the second World War; and Japan currently surpasses France technologically. We lacked resources to make such discriminations.

We already pointed out differences between the French and the Japanese regarding the autonomy of smart clothes: the former request full user control of functions, whereas the latter prefer limited artificial intelligence (Figure 7.6). To evaluate the statistical significance of this result and of other visible differences, we used t-tests for unpaired samples (Figures 7.7 and 7.8), and found significant cultural differences ($p < 0.04$) for the acceptance of enhanced garments that disclose the wearer's emotions to surrounding people, monitor physiology to produce group effects during artistic or sportive events, record videos, are under full or limited control of

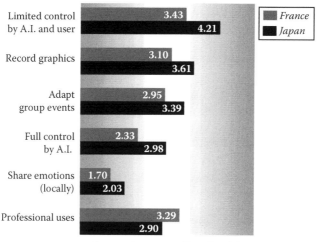

FIGURE 7.7 Items indicating a significant cultural difference ($p < 0.04$).

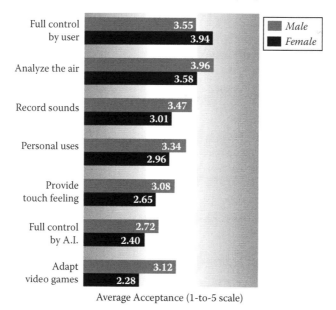

FIGURE 7.8 Items indicating a significant gender difference ($p < 0.04$).

an artificial agent, or are used for professional purposes. Similarly, we found that males accept smart clothes significantly more than females, which is consistent with the established literature in sociology. Although not a statistically significant result, readers may note with interest that females score higher than males only for a feature that restricts technology: full user control of smart clothes by wearers.

Confirming and explaining these cultural and gender differences requires additional studies, which may be guided by the two following hypotheses. Our first hypothesis is that religion guides reactions regarding autonomy: Christianity, which shapes French values, professes that creating artificial beings is evil and prohibited by God, whereas Shinto, which shapes Japanese values, suggests that objects may "naturally" be alive and have a soul whether it is a rock, river, tree, sword, toy … or computer. If confirmed, this hypothesis could help designers and researchers to easily evaluate the viability of diverse *advanced* smart clothing concepts and designs. Our second hypothesis is that the acceptance of critical features is correlated with the acceptance of, and familiarity with, other information technologies: the better acceptance of autonomous clothes and of emotional disclosure in Japan compared to France may correlate with higher technological use, higher frequency of replacement for mobile devices, and faster adoption of novel services in Japan. Descriptive rather than explanatory, this second hypothesis is weaker than the first one but, if confirmed, may lead to the replacement of complex, long, and costly social investigations by relatively simple, fast, and inexpensive analyses of technological trends.

7.5   A LIVELY HUMANITY REQUIRES DIFFERENTIATED SEEDS AS IT GROWS AND EVOLVES

Humanistic needs should be considered not only in the abstract but also in meaningful contexts, or else their seeds would remain sterile: we need seeds adapted to one's age, culture, gender, environment, experiences, and abilities. Introducing these aspects would suffice to write a full book so we will focus only on one: age. We selected this aspect because it is usually neglected, although human growth and decline induce design complexities, and although young adults will certainly request smart clothes for their children and for their own parents. Besides, designers and researchers in smart clothing should be aware of its specificities as population structures change worldwide and as governments strive to support elders in rich countries. Hereafter, we will consider the nature and features of services potentially useful to, and usable by, healthy wearers depending on their age, as well as the potential for life-long uses of smart clothing. We expect readers to understand by the end of this section that the implications of age are far reaching, even when considering only healthy cases, and that our choice was not guided by a lack of interest in disabilities.

Young people (0 to 20 years) and older adults (from 60 years) undergo important cognitive and bodily changes by phases due to phylogeny (development due to our human nature) and continuously due to ontogeny (development due to our personal history). Specialists may predict growth from birth to the end of adolescence because species-wide genetic schemes reliably guide human development during youth. However, they may not predict the decline of elders because ontogeny induces high inter- and intra-individual variability. Laymen may imagine that elders have reduced eyesight but, although the *trend* is correct, some people keep excellent eyesight until they pass away. Besides, decline may be irregular, plateauing for several years then accelerating, etc. Consequently, the worldwide young population is homogeneous whereas the old population is heterogeneous, with problems emerging from concurrent disabilities. Young and old people have different problems because decline is not a simple regression, but smart clothes useful to and usable by both groups are strikingly similar.

In 2009, smart clothes solving specific age issues are still rare due to a lack of psycho-social focus, to technological barriers, and to real-world complexity. Notable realizations include Lifebelt (Bougia, Karvounis, and Fotiadis 2007), a garment monitoring fetal and maternal vital signs during pregnancy, and Mamagoose (Verhaert 2007), a pajama designed to prevent sudden infant death syndrome. A few wearable systems like Dog@watch (WIN 2005), a wrist-worn alarm and location device for kids, could be embedded in smart clothes, but they are a far cry from what smart clothes could do with sensors covering the whole body.

7.5.1 CHILDREN'S GROWTH MAY BE ENRICHED BY SMART CLOTHES FOR PREDICTABLE DEVELOPMENT

Following human genetic schemes, young people worldwide undergo important predictable bodily and cognitive changes; nowadays, the former are well known and understood but the latter are still elusive. We sketch hereafter the potential benefits of smart clothing in relation to physical growth, perception, physiology, feelings, and cognition.

Physical growth notably depends on the environment and nutrition, and provides a convenient yardstick for global growth; it can easily be assessed until 20 years with reference charts and tables by the National Center for Health Statistics (2007) and by the World Health Organization (2007). Parents and doctors could better track children's physical growth and manage deviations with smart clothes monitoring body shapes, physical activities, and intakes thanks to stretch and movement sensors as well as cameras or radio-frequency readers. On the contrary, perception is rarely and shallowly evaluated because human sensory systems are mostly complete at birth (e.g., the vestibular system), although some elements such as the visual system mature until puberty (Brecelj 2003). Smart clothes may also strengthen perceptual skills by enabling novel games or stimulating the whole body, though specialists should first establish effective strategies and identify adverse effects.

Children's physiology differs from adults'; for example, kids are more affected by heat and cold stress (McArdle, Katch, and Katch 2007, p. 659), and adolescents require markedly more sleep (Dement and Vaughan 1999, p. 116). Smart clothes could monitor a child's body and environment to regulate his or her temperature when required, but existing sensors fail to provide inner-body data, and algorithms are still unsatisfactory. Similarly, smart clothes could monitor sleep patterns, habits, and environments (e.g., exposure to light disturbing the biological clock) to alert and advise, but appropriate combinations of sensors and algorithms are missing, notably because the importance of sleep is underestimated. Besides, diverse enhanced garments may positively influence physiology or survival, for example, by detecting allergens or pollution that may affect growth, or falls on the ground or in water, especially for infants wildly exploring and interacting with their environments. Unfortunately, specialists have apparently barely researched such garments, and pollution sensors are still too cumbersome.

In a less visible fashion, emotions also develop year after year, with a critical influence of familial contacts during infancy. Nowadays numerous children possess permanent links with their family thanks to cellular phones, suggesting our tools now suffice even for stressful separations. However, smart clothes may bring a few innovations to support distant relationships, for example, tactile stimulations mirroring a motherly hug.

Because readers probably know less about cognitive than bodily changes, we detail our following explanations more than the previous ones. Young people interpret reality with models that generate systematic errors until the accumulated and combined discrepancies lead to the adoption of better models, marking phases identified by Jean Piaget (1923) and confirmed by his followers worldwide during the following decades: the sensorimotor, preoperational, concrete operational, and formal operational phases (Table 7.2). Phase changes occur in a specific order but not at specific ages, although landmarks are typically 2, 7, and 11 years old. We briefly discuss hereafter the *formation of concepts*, *egocentrism*, *memory*, and *world views*, all related issues of particular interest for smart clothing.

TABLE 7.2
Characteristics of Young People According to Developmental Phases

Name of the Stage	Typical Age Range	Main Physical and Psychological Characteristics
• Sensorimotor	• From birth to 2 years	• Experience through senses and movements
		• Learning of object permanence
• Preoperational	• 2 to 7 years	• Acquisition of motor skills
		• Animism
		• Centration[a]
		• Classification of objects
		• Egocentrism[b]
		• Use of symbols and words
• Concrete operational	• 7 to 11 years	• End of centration
		• End of egocentrism
		• Logical thinking about concrete events
• Formal operational	• Above 11 years	• Abstract reasoning

[a] Centration: seeing problems from a single angle.
[b] Egocentrism: seeing things only from one's own point of view.

Humans develop their notion of concepts during youth by consecutively establishing syncretic heaps, complexes, and true concepts (Vygotsky 1986). Syncretic heaps look chaotic to adults, but complexes are mainly confusing because they rely on concrete factual bonds rather than abstract logical bonds. Emerging during adolescence, true concepts coexist with, rather than replace, syncretic heaps and complexes. Consequently, children should experience interface menus, explanations, and links between data as more or less natural depending on their age, which should affect general interactions with smart clothes, learning in context, and life-logging services.

During the preoperational phase, children barely recognize the difference between themselves and others (egocentrism), and frequently share their thoughts as a monologue (egocentric speech) during activities. Although apparently promising for context-aware clothes, recording and processing egocentric speech is of doubtful value because egocentric kids assume that bystanders understand what they talk about, leaving much to the imagination. Besides, egocentric speech is depressed when alone (Vygotsky 1986, pp. 233–4).

Memorial abilities correlate with age and rely on knowledge (Ornstein, Haden, and Hedrick 2004), suggesting direct uses for smart clothes. For example, life-logging clothes may provide at a later age memories otherwise unavailable because young children just retain generic temporal sequences, not event-specific information (McCormack and Hoerl 1999). Children may also better fix memories if smart clothes provide information or channel attention towards elements required to understand events, based on the child's abilities and knowledge. Although the *ubiquitous computing* community has much researched context awareness for a decade, such systems are apparently still missing.

Finally, stimulated and enlightened by objects, children develop their world views and modify their mind and behaviors under the great influence of natural and artificial things (Vygotsky 1986, pp. 39–40). Children also commonly express animism towards computers and robots, considering them as alive and intelligent (Turkle 1984). We can expect smart clothes to magnify this phenomenon as they may continuously be present and exhibit complex behaviors thanks to sensors and actuators, implying a great potential for human growth as well as a great responsibility for designers.

Services particularly useful to young people are summarized in Figure 7.9, and critical design elements are presented in Figure 7.10.

7.5.2 ELDERS' DECLINE MAY BE SMOOTHED BY DIVERSE SMART CLOTHES FOR UNIQUE CONDITIONS

Influenced by their personal history, older people worldwide undergo important but irregular bodily and cognitive changes that frequently combine to create unique conditions and difficulties in everyday life. We sketch hereafter the potential benefits of smart clothing in relation to *simple trends* of decline in physical abilities, perception, physiology, feelings, and cognition, but readers should keep in mind that these problems may interact, leading to unique situations and needs.

Children

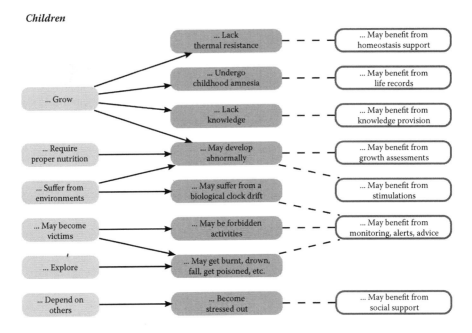

FIGURE 7.9 Services particularly useful to young people.

Elders' physical abilities decline unequally but, usually, tiredness and fatigue augment, mobility decreases, and manipulations toughen due to arthritis, less flexible limbs, increased postural sway, blurry movements, decreased strength, loss of muscle control, and longer response times (Haigh 1993; Hawthorn 2000; Morgan Morris 1994). Elders may continuously track their physical condition with sensor-embedded garments, potentially maintain their abilities with dedicated games, and frequently inform their caretakers about their health and lifestyles. The usefulness of these services should not be underestimated, especially in countries where older people live alone because, e.g., falls are a critical risk (Panel on Falls Prevention 2001; Blythe, Monk, and Doughty 2005) even for seemingly minor accidents, as ambulance crews can explain well: "[Older people's] skin will just tear, it's like tissue paper. It loses elasticity as you get older, the slightest knock can tear it" (Blythe, Monk, and Doughty 2005). Wearable computing specialists have obtained encouraging results for fall detection, but users appear unsatisfied by the form factors, leading to disuse, and specialists spend little energy to create smart garments instead.

Besides health services, designers may also personalize smart clothes with data about fatigue and abilities to increase the efficiency and comfort of whole-body interactions. These interactions, as well as more local ones, also depend on perception, which declines due to the degradation of sensory organs and associated processes; smart clothes currently exploit vision and hearing but touch may soon follow. Elders potentially suffer from diverse problems related to vision and hearing (Haigh 1993; Hawthorn 2000; Morgan Morris 1994). With a lower ability to discriminate details and distinguish between light and dark, elders may need more light to read and perceive their environment in general; they may also become more sensitive to glare

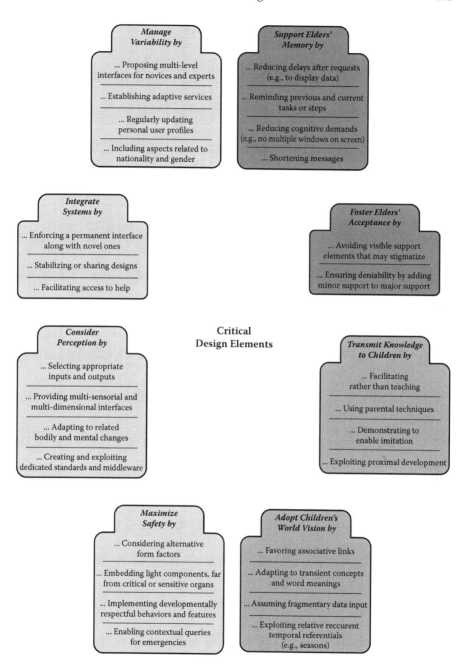

FIGURE 7.10   Critical design elements from birth to old age.

and changes in brightness, suffer from a narrower visual field and reduced depth perception, discriminate fewer colors, focus on objects and perceive rapid objects with more difficulty, and localize objects more slowly. Besides, hardened eardrums may impair the detection of sounds, particularly high frequencies, leading to ignored alarms, missed phone calls, and misunderstandings, especially in women. Finally, elders may speak less distinctly due to strokes, motor impairments, or a lower ability to hear and correct themselves; they may hesitate and restart phrases more frequently, and speak more slowly. Elders may thus particularly benefit from clothes that help them communicate, control their environments (e.g., lighting), and detect unperceived environmental dangers.

Compared to young adults, elders have fewer regular physiological activities and maintain homeostasis less reliably, notably in regard to body temperature, blood sugar levels, sleep cycles, and the control of fluids. As a result, cardiovascular problems, heat strokes, hypothermia, diabetes, falls, drowning, sleep apnea, and dementia become more common. The effects of normal aging are typically magnified or hastened as a side effect of increasing drug intakes and of drug combinations, a common occurrence as doctors frequently ignore the full list of drugs their elder patients take. Therefore, elders can benefit from smart clothes that monitor the wearer's whole body and the environment with diverse sensors, alert caretakers, support homeostasis, e.g., by warming up or cooling down the body, and protect wearers according to events by, e.g., inflating during a fall to absorb shocks. Enhanced pendants already exist to identify falls and contact caretakers, but elders frequently forget them when they go to the restroom at night or wake up in the morning, a problem that would vanish with daytime smart garments or pajamas. Besides, diverse enhanced garments may limit negative influences on physiology, for example, by detecting allergens or pollution.

Elders also indirectly suffer from physical, sensory, and physiological decline as they become more isolated due to their own loss of mobility, as well as to the loss of mobility and progressive disappearance of their old close friends. Smart clothes are not usually conceived to increase mobility, but exoskeletons like HAL (Sankai Laboratory 2007) could be integrated or adapted with smart materials to fulfill this goal, with the caveat that energy would be a major obstacle. Besides, smart clothes could provide permanent artificial companions to wearers, though researchers should investigate their effects on the human psyche, especially for elders affected by dementia.

Elders globally preserve their intellectual abilities until old age (Hawthorn 2000), especially for routine or narrow but deep expertise like chess (for champions), but are easily impaired by disturbances and simultaneous activities (Haigh 1993; Morgan Morris 1994), notably due to a marked decrease in their ability to process information (work memory, spatial memory, content recall), although holding information (short-term memory, prospective memory, recognition of familiar objects) is only moderately affected by age (Hawthorn 2000). Healthy elders should therefore benefit more from advanced smart clothes meant to deal with infrequent events and to process intensive activities than from basic smart clothes supporting common or simple tasks; elders may of course also benefit from smart clothes stimulating them to slow down or reverse their decline. For example, serious games based on the five senses and cognitive challenges may preserve or redevelop muscles, perception, and memory. However, the interfaces should remain undemanding, avoiding long messages as

Older Adults

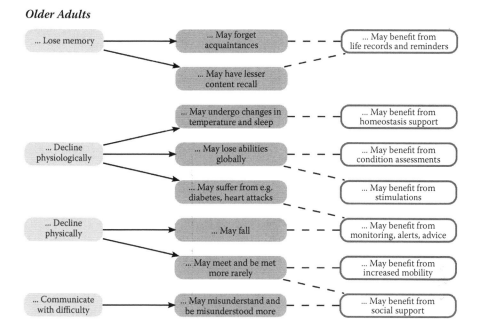

FIGURE 7.11 Services particularly useful to older adults.

well as simultaneous text and speech (Zajicek and Morrissey 2003). As long as they find ways to bypass their memorial limits, elders can acquire skills but they have difficulty acquiring automated responses (Hawthorn 2000), which means they need more cognitive resources to manipulate unfamiliar tools such as … smart clothes.

Finally, elders typically wish to maintain or regain their independence and freedom to take risks (Blythe, Monk, and Doughty 2005), and may thus adopt systems, once convinced of their usefulness. However, elders may reject these systems to avoid looking dependent or old (Morgan Morris 1994), a common explanation for the disuse of easy-to-notice pendants alerting caretakers after falls. Elders should accept smart clothes more than these pendants because they may be designed to look like normal clothes, and may provide multiple services, allowing wearers to deny they use a given garment for a given service; relatives may not be fooled, but elders may at least save face.

Services particularly useful to older adults are summarized in Figure 7.11, and critical design elements are presented in Figure 7.10.

7.6 HUMANISM MAY SUCCESSFULLY SEED VISIONS OF SMART CLOTHING FOR WORLD CITIZENS

Founded on sound psychological research but globally unexploited in smart clothing so far, humanistic needs can provide meaningful and fertile seeds to envision smart clothes for the general public worldwide. Besides, the reactions of French and Japanese citizens suggest that humanistic seeds are of immediate practical relevance to plan the creation of smart clothes or otherwise explain their success or failure in various cultures and environments. Applicable as soon as a human is involved, these highly conceptual seeds can reliably guide experts towards superior designs of specific services as well as towards superior selections of services for specific users. Importantly, humanistic bases are valid independently of moments and innovations; specialists just have to carefully choose and remember the assumptions they associate with humanistic seeds. Hereafter, we present our current vision of smart clothing then discuss related core aspects to guide stakeholders.

7.6.1 WHAT PERTINENT VISION DO WE PROPOSE?

We envision smart clothes that gratify men's and women's humanistic needs from birth to old age, respect individuals and human diversity worldwide, continuously comply with diverse lifestyles and environments, and exist in harmony with nature. Readers may more easily remember the characteristics of this vision as *universal gratification*, *universal access*, *universal respect*, *universal compliance*, and *universal sustainability*. These concepts emerged from our evaluation of humanistic needs as seeds for smart clothing, focused on individuals rather than groups, with diverse influences highlighted by our theoretical and practical investigations. Besides, these concepts naturally extend the standard qualities of garments: protection from given environments, reassurance, enhancement of relationships in given cultures, reinforcement of esteem, and empowerment for everyday activities. In our vision, smart garments deeply relate to the benefits of fairness and to the costs of wasted human potential.

By universal gratification, we mean that ideally anybody could benefit from smart clothes that gratify their own specific needs. The essence of humanistic needs is universal, but we consider here their unique expression for a given person at a given moment; it matters because one's needs change with age, and change differently according to one's personal history. Inventors may create smart clothes for diverse goals but only if they perceive needs, possess adequate knowledge, and are motivated. As a consequence of universal gratification, smart clothes may monitor the condition of pregnant women and of their fetus, reassure children at night, cool down adult hikers, absorb shocks of falling elders, and help foreigners communicate.

Universal access is a standard concept in technology; here it means that ideally anybody could efficiently use services provided by smart clothes. It matters because one's physical and mental abilities may vary widely, particularly due to age, (dis-) abilities, and training. Thanks to various sensors and actuators covering the body, smart clothes may allow flexible interactions but tailoring may initially be expensive. As a consequence of universal access, children who cannot read may use iconic menus; the blind may follow tactile guidance; and geeks may master their garments with cryptic commands.

By universal respect, we mean that ideally everybody would have access to smart clothes that respect their values and traditions. It matters because everybody is permanently and continuously influenced by one or several cultures, frequently also by religion, and because non-respect can decrease quality of life as well as lead to opposition movements. Companies may tailor smart clothes differently depending on countries, or may design the garments to automatically adapt their behavior to the values and traditions of their wearer. As a consequence of universal respect, smart clothes may be semi-autonomous when worn by Japanese users but be under full manual control when worn by French users.

By universal compliance, we mean that ideally, smart clothes could be properly used in any context. It matters because lifestyles and environments vary so much worldwide. As with standard clothes, we do not really expect similar comfort and functionality in a temperate city, in the Sahara, or at the top of Mount Everest, but we expect smart clothes to remain efficient in a reasonable range of conditions, and we expect appropriate smart clothes for each environment in which humans already dwell. As a consequence of universal compliance, the garments of Korean disc jockeys may light up at night in Seoul, whereas those of British gentlemen may remain dim in London.

Finally, by universal sustainability we mean that ideally, smart clothes would have no negative impact on nature. It matters because human activities already disrupt natural networks worldwide and reduce available resources. Ideally, then, smart clothes would generate renewable energy, would be recyclable or biodegradable, and would reduce the environmental footprint of humanity. At the moment, specialists ignore how to deal with energy and waste, but we may quickly start reducing the footprint of humanity; for example, smart clothes may increase environmental awareness and thus lead to behavioral changes, or may reduce transportation by replacing face-to-face meetings with high quality telepresence thanks to whole-body sensing and stimulation.

7.6.2   WHAT FACTORS SEEM CRITICAL TO REALIZING THIS VISION?

To identify potential locks for the realization of this vision, we first consider general issues in smart clothing, then the specificities of humanistic clothes, the management of diversity, the existence of technological ecologies, the features proposed by Mark Weiser, and finally, acceptance.

Practically, smart clothes are garments that contain electronic components. As we focus here on the general public, designers should ensure that the clothes are comfortable in everyday life and functional for common activities; the smart elements should therefore be small or flexible, well integrated, resistant to daily wear, and washable. As indicated elsewhere in this book, researchers have already created textiles that merge usual fibers and electrically conductive fibers, and that remain in good shape after standard abrasion and washing tests. Therefore, basic issues of comfort and functionality remain only for the integration of sensors and actuators, with predictable difficulties to ensure good properties and safety in cold, hot, humid, or dusty environments. Besides, enhanced textiles should remain pleasant to see and touch, because most people and fashion designers care about image; in this respect, nanotechnologies may be extremely useful, allowing additional miniaturization and more precise engineering of materials. The major lock for the creation of smart clothes is currently energy, and no solution is in sight; as a consequence, designers should keep the consumption of electronic components at a minimum, and researchers should investigate more thoroughly energy production by garments, maybe using photovoltaic technologies.

Besides, humanistic clothes raise issues related to physiology, safety, belonging, and esteem. Most related garments support health, and the integration of sensors seems difficult, especially since reliability is critical for, e.g., cardiac patients. In addition, novel sensors and algorithms should be developed: sensors to detect allergens and pollution are oversized, systems can conveniently measure skin temperature but not inner-body temperature, and we lack algorithms to identify sleep troubles from sleepers' movements and sounds. Besides, humanistic clothes should be physically and mentally safe, notably taking into account the risks in case of car accident, fall, or malfunction, and the possible increases in stress, losses of abilities, or losses of self-esteem due to, e.g., intelligent garments replacing its wearer for crucial tasks. To enhance physical safety, designers may change form factors, or the properties, weight, and location of components depending on wearers' assumed characteristics, based on personal experience and on the large body of knowledge available. For example, garments should not warm up children much, and tactile interfaces should be particularly gentle with elders to avoid short-term as well as long-term adverse effects. Designers will have difficulties avoiding negative side effects on mental abilities and ways of thinking because much knowledge is lacking; for example, we ignore how artificial vibrations (enhancing communication or entertainment) felt daily for years may affect brain development, we ignore how automatic adaptations of interfaces (more convenient but less challenging) may affect abilities in old age, and we ignore how the youngest and oldest will change their world views when supported by seemingly intelligent/alive garments. Finally, to ease wearers' concerns over cyber-security, designers may focus on open designs, enable secure update systems, favor anonymity or pseudo-anonymity for both wearers and hardware, and allow non-vital systems to be turned off on demand.

Beyond universal aspects, diversity is inevitable due to high human inter- and intra-variability in needs and abilities, which notably fluctuate markedly during growth and decline, sometimes over months as parents can notice, and sometimes over hours as diabetics and elders know. Designers should think about service issues on a case by case basis, and may need novel algorithms to implement reliable interfaces based on standard techniques such as multi-level interfaces, artificial intelligence, or adaptation to users' states, maybe exploiting up-to-date profile data about users. A specific issue for the life-long use of smart clothes is that personal data, probably based on sensor acquisition, should be reusable when people purchase clothes based on new technologies and/or when the characteristics of one's body change, which may require the development of well-thought-out standards. Finally, designers will certainly have difficulties managing cultural diversity, and we believe that open/free dedicated databases should be created accordingly, by gathering materials from psychologists, sociologists, and computer scientists, and feedback from wearers.

According to our vision, smart clothes may form a specific technological ecology, with old and new technologies cohabiting. First, smart clothes may be extremely diverse because they gratify the specific needs of numerous wearers while respecting their values and while operating in various environments, and because people may keep smart clothes as long as common clothes; this highlights the importance of hardware, software, and data standards for maintenance, upgrades, and interoperability. Besides, designers should favor stable or shared designs, standard metaphors, and natural interactions to facilitate the use of new systems in everyday life. At the moment, we ignore how to implement such designs in a complex and dynamic ecology of devices and services on a world scale; smart clothes appear particularly tricky because they do not rely on screens like most other information systems. Second, garments may collaborate with other garments, accessories, and implants on the same person; for example, bras may inform about cardiac activity, jackets about air temperature, implants about body temperature, and bags about air quality, requiring energy-efficient secure communication techniques. Third, smart clothes may depend on outside infrastructures, for example, wireless infrastructures or food distribution infrastructures that ensure the addition of descriptive RFID tags in all food packages. Finally, technological ecologies may emerge from, and evolve according to, various corporate and legal practices.

Mark Weiser proposed context awareness and invisibility as two fundamental concepts for ubiquitous computing, but they do not appear critical in smart clothing, as they did not naturally emerge from our analyses. Of course, context awareness may be extremely useful for specific services, for example, to help children understand their environment and accordingly fix their memories, or to support elders at risk due to reduced attention and perception. In any case, designers should beware of context awareness because its theoretical foundations are shaky: so far approaches are non-scalable, too restrictive, or simply too optimistic for a complex dynamic world; the situation is even worse for affective awareness as psychologists even disagree about the feasibility of evaluating emotions with physiological sensors in everyday life. Instead of generic context awareness, we would rather promote the modeling of human needs and the creation of wearer profiles, which would provide a smart garment with deep knowledge of its wearer. Although potentially useful for adaptive systems and for advisory systems, the modeling of human needs seems unexplored. As for invisibility, it seems antithetic to humanistic needs as it may make wearers feel insecure: people may wonder if a conversation is being recorded, if a critical service is available, etc.; such worries may lead to the rejection of smart clothes, suggesting that invisibility should be avoided or well balanced.

Finally, the acceptance of smart clothes itself is critical, and requires further investigations. These investigations should involve children, young adults, and elders; men and women; and people from various backgrounds. To gather valid data, several methods may be used, for example, scenario-based feedback with children and dilemma-based feedback with elders. The reactions of early adopters should be studied, maybe with anthropological methods, and specialists should carefully study adoption by elders as they may reject a garment if they feel it makes them look old or dependent. Readers may think that solving this problem is trivial but real-world deployments of ubiquitous systems have proved otherwise.

7.7 PERSPECTIVES

Based on the sound psychological works of Abraham Maslow, humanism may successfully seed visions of smart clothing worldwide, with predictable benefits for individuals and societies within 5 to 10 years. Humanistic needs seem to be good seeds because they may hasten the creation of smart clothes significantly improving quality of life, and lead to the creation of garments that would never have appeared otherwise due to an excessive focus of stakeholders on the dominant vision of ubiquitous computing. Smart clothes at hand cover so few true needs that human potential would be dramatically wasted without a reorientation and integration of investigations by specialists from art, clothing, computer science, electronics, fashion, medicine, psychology, and sociology fields. Although these generic seeds already provide tools to predict and explain the reactions of the general public towards smart clothing concepts, they should be differentiated to better encompass the uniqueness of our lives, guiding experts towards superior selections and superior designs of services. In the vision we propose, smart clothes thus gratify men's and women's humanistic needs from birth to old age, respect individuals and human diversity worldwide, continuously comply with diverse lifestyles and environments, and exist in harmony with nature. We accordingly expect smart clothes to become more useful and desirable while improving the general public's quality of life worldwide.

During the next 6 years, we shall check, deepen, and extend our hypotheses with bigger-scale studies in South Korea, Japan, France, and Australia using better sampling and various methods to study variations due to human diversity and environments, for example, using drawings and scenarios with children. We shall also model individuals' reactions to smart clothing for everyday life, incorporating the perspectives of other major psychologists and sociologists such as Erving Goffmann and Carl Rogers. Depending on opportunities, we may finally consider aspects related to groups in addition to individuals, and update our vision accordingly.

Looking broadly at the theoretical aspects, we may ask, is the influence of Maslow's needs in smart clothing really universal? How are these needs weighted against each other? What would be good concepts of clothes gratifying belonging or esteem needs? Can smart clothes gratify self-actualization needs, and if so, how? Does religion guide reactions regarding the autonomy of smart clothes? How shall we adapt smart clothes to diverse cultures? How may we best develop environment-friendly smart clothes?

Finally, looking broadly at the practical aspects, we may ask, what images of emerging information technologies are shared worldwide? How should governments promote humanistic seeds, e.g., legally or financially? Is the acceptance of critical features correlated with the acceptance of, and familiarity with, other information technologies? What would be effective strategies to strengthen perceptual skills with smart clothes? What smart clothes should we design first, considering the energy lock? What is the influence of lack of experience with actual systems on adoption?

REFERENCES

Amft, O., H. Junker, and G. Troster. 2005 Detection of eating and drinking arm gestures using inertial body-worn sensors. Paper presented at International Symposium on Wearable Computers, 160–63.

Bell, G., M. Blythe, and P. Sengers. 2005. Making by making strange: Defamiliarization and the design of domestic technologies. *ACM Transactions on Computer-Human Interaction* 12 (2): 149–73.

Bell, G., and P. Dourish. 2007. Yesterday's tomorrows: Notes on ubiquitous computing's dominant vision. *Personal and Ubiquitous Computing* 11 (2): 133–43.

Blythe, M., A. Monk, and K. Doughty. 2005. Socially dependable design: The challenge of ageing populations for HCI. *Interacting with Computers* 17 (6): 672–89.

Borovoy, R., F. Martin, S. Vemuri, M. Resnick, B. Silverman, and C. Hancock. 1998. Meme tags and community mirrors: Moving from conferences to collaboration. Paper presented at ACM conference on Computer Supported Cooperative Work, 159–68.

Bougia, P., E. Karvounis, and D. Fotiadis. I. 2007 Smart medical textiles for monitoring pregnancy (chapter 10). In *Smart textiles for medicine and healthcare—Materials, systems and applications.* Cambridge: Woodhead Publishing in Textiles.

Brecelj, J. 2003. From immature to mature pattern ERG and VEP. *Documenta Ophthalmologica* 107 (3): 215–24.

Dement, W. 1976. *Some must watch while some must sleep.* New York: Norton and Co. Inc.

Dement, W., and C. Vaughan. 1999 *The promise of sleep.* New York: Dell Trade Paperback.

Edmison, J., D. Lehn, M. Jones, and T. Martin. 2005. User's perceptions of an automatic activity diary for medical annotation and analysis. Paper presented at International Symposium on Wearable Computers, 192–93.

Falk, J. S., and Bjork. 1999. The BubbleBadge: A wearable public display. Paper presented at CHI, 318–19.

Haigh, R. 1993. The ageing process: A challenge for design. *Applied Ergonomics* 24 (1): 9–14.

Hawthorn, D. 2000. Possible implications of aging for interface designers. *Interacting with Computers* 12 5:507–28.

Healey, J., and R. W. Picard. 1998. StartleCam: A cybernetic wearable camera. Paper presented at International Symposium on Wearable Computers, 42.

Jafari, R., F. Dabiri, P. Brisk, and M. Sarrafzadeh. 2005. Adaptive and fault tolerant medical vest for life-critical medical monitoring. Paper presented at ACM Symposium on Applied Computing, 272–79.

Jiang, X., J. I. Hong, L. A. Takayama, and J. A. Landay. 2004. Ubiquitous computing for firefighters: Field studies and prototypes of large displays for incident command. Paper presented at CHI, 679–86.

Kanis, M., N. Winters, S. Agamanolis, A. Gavin, and C. Cullinan. 2005. Toward wearable social networking with iBand. Paper presented at CHI: Extended Abstracts on Human Factors in Computing Systems, 1521–24.

Kaur, S., R. Clark, P. Walsh, S. Arnold, R. Colvileand, and M. Nieuwenhuijsen. 2006. Exposure visualisation of ultrafine particle counts in a transport microenvironment. *Atmospheric Environment* 40 (2): 386–98.

Kawahara, Y., C. Sugimoto, S. Arimitsu, T. Itao, A. Morandini, H. Morikawa, and T. Aoyama. 2005. Context inference techniques for a wearable exercise support system. Paper presented at Siggraph, 91.

Lifewatcher. 2008. http://www.lifewatcher.com/.

Maslow, A. 1987. *Motivation and personality* (3rd Ed.). Addison Wesley Longman.

McArdle, W., F. Katch, and V. Katch. 2007. *Exercise physiology: Energy, nutrition, and human performance* (6th Ed.). Baltimore: Lippincott Williams & Wilkins.

McCormack, T., and C. Hoerl. 1999. Memory and temporal perspective: The role of temporal frameworks in memory development. *Developmental Review* 19 (1): 154–82.

Morgan Morris, J. 1994. User interface design for older adults. *Interacting with Computers* 6 (4): 373–93.

National Center for Health Statistics. 2007. NCHS Growth Charts. http://www.cdc.gov/growthcharts/.

Ornstein, P., C. Haden, and A. Hedrick. 2004. Learning to remember: Social-communicative exchanges and the development of children's memory skills. *Developmental Review* 24 (4): 374–95.

Panel on Falls Prevention. 2001. Guideline for the prevention of falls in older persons. *Journal of the American Geriatrics Society* 49 (5): 664–72.

Piaget, J. 1923. *Le language et la pensée chez l'enfant*. Paris: Delachaux et Niestlé.

Picard, R., and J. Scheirer. 2001. The Galvactivator: A glove that senses and communicates skin conductivity. Paper presented at International Conference on HCI.

Rantanen, J., J. Impio, T. Karinsalo, M. Malmivaara, A. Reho, M. Tasanen, and J. Vanhala. 2002. Smart clothing prototype for the arctic environment. *Personal and Ubiquitous Computing* 6 (1): 3–16.

Rivers, T. 2008. Technology's role in the confusion of needs and wants. *Technology in Society* 30 (2): 104–9.

Sankai Laboratory (Tsukuba, Japan). 2007. Robot Suit HAL—Hybrid Assistive Limb. http://sanlab.kz.tsukuba.ac.jp.

Sensatex. 2005. http://www.sensatex.com.

Sleeptracker. 2005. http://www.sleeptracker.com/.

Smart Second Skin. 2008. http://www.smartsecondskin.com/.

Tierney, M., J. Tamada, R. Potts, L. Jovanovic, S. Garg, and Cygnus Research Team. 2001. Clinical evaluation of the GlucoWatch biographer: A continual, non-invasive glucose monitor for patients with diabetes. *Biosensors and Bioelectronics* 16 (9–12): 621–29.

Turkle, S. 1984. *The second self: Computers and the human spirit*. New York: Simon and Schuster.

United States Army. 2005. Soldiers testing cooling vests in Iraq. http://www.militaryconnections.com/news_story.cfm?textnewsid=1791.

Verhaert. 2007. Mamagoose pyjamas. http://www.verhaert.com/cms/images/stories/pdf_in_text/verhaert_mamagoose.pdf.

Vivometrics. 2005. Lifeshirt. http://www.vivometrics.com.

Vygotsky, L. 1986. *Thought and language*, Ed. Alex Kozulin. Cambridge: MIT Press.

Weiser, M. 1991. The computer for the 21st century. *Scientific American* 265 (3): 66–75.

Weiser, M., and J. Brown. 1997. Beyond calculation: The next fifty years of computing (chapter 6). In *The coming age of calm technology*, 75–85.

WIN. 2005. Aware wear proposal by Wearable Fashion Project (in Japanese). *Nature Interface*.

World Health Organization (WHO). 2007. The WHO Child Growth Standards. http://www.who.int/childgrowth/en/.

Zajicek, M., and W. Morrissey. 2003. Multimodality and interactional differences in older adults. *Universal Access in the Information Society* 2 (2): 125–33.

Zieniewicz, M. J., D. C. Johnson, D. C. Wong, and J. D. Flatt. 2002. The evolution of army wearable computers. *IEEE Pervasive Computing* 1 (4): 30–40.

8 Shape Memory Material

Chang Gi Cho

CONTENTS

8.1 INTRODUCTION

Shape memory materials (SMMs) are a family of materials that can change their shapes from temporarily deformed shapes to their programmed original shapes. The shape recovery is usually activated by the surrounding temperature, but other stimuli such as electric field, magnetic field, pH, UV light, specific chemical, or any other stress can also trigger the shape recovery behavior. Depending on the extent of recovering force, and characteristics of recovering mechanism,

SMMs can be applied to control or tune many technical parameters in smart material systems. Shape memory material is a type of smart material whose properties can be changed in a controlled fashion by external stimuli. Shape, position, strain, stiffness, damping, friction, vapor permeability, and surface tension are some examples that can be manipulated through smart material systems using SMMs.

SMMs are either inorganic or organic materials based on their constituents. Inorganic materials include alloys and ceramics, and organic materials include polymers and gels. Among them, shape memory alloy (SMA) is the one that has the longest history and the widest commercial application areas. The biomedical area especially has lots of successful application. Other extensively researched areas for shape memory alloy include the space and aircraft industry, automotive industry, micro electromechanical systems (MEMS), and the telecommunication area. Shape memory alloys and shape memory polymers are thermoresponsive materials where deformation can be induced and recovered through temperature changes. Ceramic material's history is not so long, so the material is rather at the research stage. Polymers and gels were introduced not long ago and the application area is limited in some specific areas, but the area is rapidly expanding. Compared to inorganic materials, organic materials have large recoverable deformation, easily adjustable temperature, and low cost. Therefore, shape memory polymers (SMPs) and gels are good candidates for the application where the material cost is a critical issue. But the application of gels is limited due to their poor mechanical properties (Mattila 2006a, 2006b).

8.2   SHAPE MEMORY ALLOY

Shape memory alloy (SMA) memorizes its programmed original shape and can recover its original shape from deformed temporary shape. The shape recovery is usually activated by the heating of the surrounding temperature. Together with piezoelectric ceramics, electrostrictive materials, and magnetostrictive materials, SMA is one of the smart inorganic materials that are widely used today. All these smart inorganic materials undergo at least two crystallographic phase transformations by temperature change, and show shape memory effect (Newnham 1998).

The history of SMA can be traced back to 1932 when a Swedish physicist discovered the pseudoelastic behavior of Au-Cd alloy (Otsuka and Wayman 1998). Since the discovery, many materials have been reported in the category of SMA, and two families of alloys are commercialized—one is Ni-Ti alloy and the other is Cu-Zn-Al alloy. But the most commonly used material is nitinol, nickel-titanium alloy, which was first discovered at the U.S. Navy ordnance laboratories in the 1960s, after the researchers accidentally observed that the alloy changed its shape by the touch of a burning cigarette. When the nickel-titanium composition is near equiatomic of 50:50, nitinol undergoes a solid-state phase transformation in the temperature range of –200°C to +100°C. Small compositional variations around 50:50 atomic ratio shift the phase transition temperature and make dramatic changes in the thermomechanical characteristics of the alloy (Tang et al. 1999). Alloys with slightly nickel rich compositions result in the phenomenon known as superelasticity, and it is this behavior that is utilized in the vast majority of medical applications. Addition of a third component such as Cu, Al, Fe, Cr, Co, Pt, Pd, Au, or Hf to the nitinol has been tried to modify the properties of the alloy in order to improve processability, phase stability, and mechanical stability (Fuentes, Guempel, and Strittmatter 2002).

8.2.1 CRYSTALLINE PHASE TRANSFORMATION

When nitinol is cooled down from a relatively high temperature to a low temperature, a broad phase transition occurs from partially ordered cubic or cesium chloride-like structure (austenite) to a monoclinic structure of lower symmetry (martensite). In the austenite phase, nickel and titanium atoms are arranged alternately in a crystal lattice structure (Figure 8.1a). The phase transformation to martensite is made without any long-range diffusion of atoms. Once it reaches the martensite phase, the shape of the alloy can be easily deformed by the application of an outside mechanical force, by adjusting the bond angles connected to mirror plane (twinning) (Figure 8.1c). So the connectivity between individual atoms is maintained throughout the deformation process within a limited range of maximum strain of 8%, while most other metals deform by slip or dislocation of crystal lattice. Due to this reason, the martensite phase is softer than austenite in nitinol. So the deformation can be made easily in the martensite phase and this deformation can be recovered when the phase of the alloy is returned to austenite by heating.

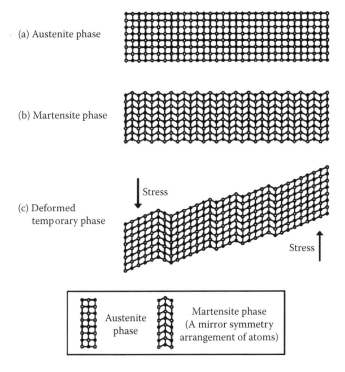

FIGURE 8.1 Schematic illustration of change of crystal lattice structure of nitinol on phase transformation and with applied stress.

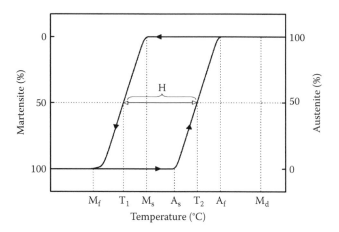

FIGURE 8.2   Phase behavior of nitinol versus temperature.

8.2.2   HYSTERESIS

As shown in Figure 8.2, at high temperature, nitinol is 100% austenite phase, and as the nitinol is cooled down, it reaches the so-called martensite start temperature (M_s). Below this temperature, the amount of austenite phase is decreased gradually and at the same time the amount of martensite phase is increased. When the nitinol is cooled down further, it passes the so-called martensite finish temperature (M_f) below which the phase of nitinol is 100% martensite. When the nitinol is heated again, the temperature versus phase profile does not follow the cooling curve, but shows hysteresis behavior. As in Figure 8.2, the nitinol needs to be heated 20°C~30°C above M_f to show the austenite start temperature (A_s). When temperature is increased, the nitinol reaches the austenite finish temperature (A_f) above which the phase is 100% austenite. Therefore, the temperatures to make 50% of martensite on cooling (T_1) and that to make 50% of austenite on heating (T_2) are different, and the difference ($T_2 - T_1$) is generally defined as hysteresis (H). As the hysteresis increases, shape memory effect is better, and the nitinol is the one that shows the biggest hysteresis among many SMAs.

Tang et al. show M_s and A_f values of nitinols with different compositions (Tang et al. 1999). The value of hysteresis becomes very close to the difference between M_s and A_f, and the use of A_f is very practical to know the real applicable temperature.

8.2.3   SUPERELASTICITY

When a metal is deformed by elongation or compression within a very limited range, it shows an elastic behavior, i.e., it returns to the original shape after the removal of the outside stress. But if the metal is deformed beyond the elastic limit, it reaches yield point in a stress-strain curve and undergoes plastic deformation, and then the shape of the metal is not restored even after the removal of the outside mechanical stress. For a metal like steel, this elastic limit is very low, such as 0.5% or 1.0% in strain. But for a *superelastic* alloy this elastic limit is more than several percent.

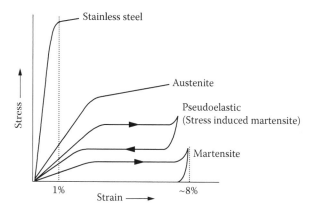

FIGURE 8.3 Typical stress-strain curves at different temperatures relative to transformation, showing martensite, pseudoelastic behavior, austenite, and stainless steel.

When the austenite phase of nitinol is under mechanical stress, the austenite phase can be transformed into martensite (stress-induced phase change), and the nitinol can be deformed to a relatively large extent due to the deformability of the martensite. After unloading of the external stress, the alloy returns to its original shape. Above the austenite finish temperature, the shape recovery process is totally reversible from relatively large deformation, so it has *superelasticity*. The stress-induced martensite (SIM) occurs until the temperature is equal or below the martensite deformation temperature (M_d) as shown in Figure 8.2, above which no superelasticity is observed. Hardness of austenite increases as the temperature increases.

Also, when the temperature of the nitinol under mechanical stress is below the austenite finish temperature ($M_f < T < A_f$), the shape recovery is not 100% but needs to heat the sample above the austenite finish temperature to become fully recovered. So the alloy under the austenite finish temperature is *pseudoelastic*. This effect can be indirectly related to a *damping* effect of a material. When the temperature of the nitinol is below the martensite finish temperature ($T < M_f$), the alloy stays at the deformed state then returns to its original shape upon heating. The terminology of *pseudoelasticity* is sometimes called *superelasticity*. The nearly horizontal (plateau) region during unloading in Figure 8.3 has been defined as the superelastic region (Chang and Nikolai 1994). The nitinol, therefore, has thermal memory effect as well as elastic memory effect.

8.2.4 ONE-WAY MEMORY AND TWO-WAY MEMORY

Nitinol's shape is set at high temperature; for example, if nitinol is heated at 400°C~500°C around 30 minutes with a certain shape, then the shape is memorized. After cooling the alloy, the shape is deformed to a temporary shape at low temperature, then the memorized shape can be restored by heating. But no shape change can occur by the subsequent cooling cycle. In order to observe another shape memory behavior the alloy needs to be deformed again. This kind of behavior is called the one-way shape memory effect or one-way memory (OWM). When an alloy memorizes both high temperature shape and low temperature shape at the same time, however, this kind of behavior is called two-way shape memory effect or two way memory (TWM). TWM looks to be very attractive in its principle, but this effect can be imparted only by several complicated training processes, and the extent of maximum strain is usually low (Chang, Vokoun, and Hu 2001). So a device with TWM is usually constructed by the combination of SMA and a mechanical spring.

8.2.5   APPLICATIONS

8.2.5.1   Biomedical Applications

Nickel titanium (nitinol) has become a material of choice for many applications in the medical device industry. Because nitinol is typically composed of approximately 50% to 55.6% nickel by weight, making small changes in the composition can change the transition temperature of the alloy significantly. For this reason, nitinol has been studied well. And the most successful application example is the *self-expanding* stent. A stent is a device that is used to prevent a natural conduit such as a blood vessel from constriction. There are two kinds of stents—one is a *balloon expanding* stent, which is made of stainless steel, and the other is a *self-expanding* stent made of nitinol. Between the two, the nitinol stent is more popular due to its smart nature: the nitinol stent has *biased stiffness* due to the stress hysteresis when the balance of forces in a stented vessel is examined. The concept is illustrated in Figure 8.4, which shows a typical stress-strain curve for nitinol, illustrating both non-linear response and hysteresis.

A stent of a given size larger than the vessel is crimped into a delivery system (from point *a* to *b* in Figure 8.4), then packaged, sterilized, and shipped. After insertion to the target site, the stent is released from the delivery system into a vessel and expanded by the body temperature until movement is stopped by impingement with the vessel (from point *b* to *c*). At this point, further expansion of the stent is prevented. But because the stent did not expand fully to its programmed original shape, it continues to exert a low outward force, so-called *chronic outward force* (COF). At the same time, it will resist recoil pressures or any other external compression forces

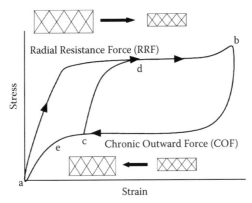

FIGURE 8.4   Stress-strain responses of a self-expanding nitinol stent.

due to forces as shown by the loading or crimping curve (from point *c* to *d*), which is substantially stiffer than the unloading direction towards *e*. These forces are termed *radial resistive forces* (RRF). In an ideal stent the RRF is designed as high as possible, and the COF is set as low as possible.

Orthodontic wire tightener made of nitinol reduces the need for adjustment and retightening of the wire. The superelastic force of the wire aligns the tooth with little additional force applied to the tooth. It is necessary to chill the wire outside the mouth and keep the wire chilled during bending and ligation (Mullins, Bagby, and Norman 1996).

Nitinol clamps and staples are used to fasten fractured bones inside the body. At body temperature the nitinol tends to shrink back to its original length, so the close contacts of fractured bones are ensured.

8.2.5.2 Clothing and Textile Applications

Joan Benoit was the women's marathon winner at the 1984 summer Olympics in Los Angeles, and she is more famous because of her brassiere made with SMA. Shortly after, the bra with SMA was commercialized, and gives a good look and comfort to its wearer. The deformed shape of SMA wire in a bra during laundering can be restored by the body temperature with wearing.

SMA wires or springs can be incorporated into clothing. The Italian company called Italian Fashion House Corpo Nove attempted the use of SMA in a designer shirt that rolls up its sleeves when the body gets warm. SMA wires are also incorporated into handbags or suitcases where the shapes are maintained by the SMA, and this increases the aesthetic appearance of personal items.

TWM SMA springs can be applied to garments for firefighters as illustrated in Figure 8.5 (Kim, Jee, Kim et al. 2008). The SMA springs are located between two

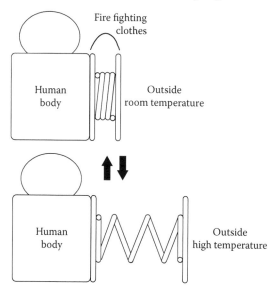

FIGURE 8.5 Schematic illustration of TWM SMA spring in fire fighting clothes.

inflammable fabric layers. At ambient temperature the garment is lightweight and thin, but the thickness becomes thicker to allow necessary insulation for protection at high temperature. The SMA springs are known to show excellent performance compared to currently existing fire fighters' jackets based on bimetal devices. Around 30% increase in protection time was reported.

8.2.5.3 Aerospace and Vehicle Applications

During the Apollo 11 mission in 1969, a self-deployable parabolic antenna was first used in the lunar lander. Assembled at 150°C and folded at room temperature to a small size, the antenna was carried to the moon where the surface temperature can reached up to 200°C by the sunlight. With the sunlight, therefore, the antenna was self-deployed.

Cryofit hydraulic couplings are manufactured as cylindrical sleeves whose diameters are slightly smaller than the tubings they are to connect. The diameter of the coupling is enlarged at low temperature and the ends of the two tubes are inserted into the coupling for connection. By heating the coupling, the diameter shrinks to hold the tube ends tightly. These couplings join hydraulic lines tightly and easily, and have been used in fighter planes since the late 1960s (Borden 1991). This technology has been commercialized by Aerofit for more than 40 years.

8.2.5.4 Housing and Safety Device Applications

SMA is applied in an emergency fire door. In case of fire, the hinge pin of the door is heated and detached from the hinge, so the door can be easily opened from outside for fire fighters. The fire extinguishing sprinkler is another example of safety device application. Generated heat by a fire activates the sprinkler valve made of SMA to spray water. A smart actuator that opens or closes windows of a greenhouse with the temperature variation is another example. At high temperature the window is opened, and at low temperature the window is closed by a smart arm made from SMA.

8.2.5.5 Other Applications of Interest

In a micro-robotic system, a joint design is a challenging task. SMA is frequently used as an artificial muscle in robotic joint systems. Because the mechanical force exerted in returning to original shape for SMA at high temperature (above A_f) is higher than the force to deform the SMA at low temperature, a contractive motion of an SMA spring can be easily activated by the temperature change. SMA is also used to detect visually an abnormal heating of power lines for electric trains.

8.3 SHAPE MEMORY POLYMER

The shape memory polymers (SMPs) are a class of intelligent or smart materials that receives great attention in a number of applications (Behl and Lendlein 2007; Fakirov 2005; Liu, Qin, and Mather 2007). Smart material is a material that can be changed significantly by an external stimulus. Some people say it is *intelligent material*. Such applications can be found in, for example, smart fabric and clothing (Hu 2002; Inoue et al. 1997; Jeong, Ahn, and Kim 2000; Mondal and Hu 2006; Tobushi et al. 1996), heat shrinkable tubes for electronics or films for packaging (Charlesby 1960), self-deployable sun sails in spacecraft (Campbell 2005), self-disassembling mobile phones (Bhamra and Hon 2004),[*] intelligent medical devices (Wache et al. 2003), or implants for minimally invasive surgery (Lendlein and Langer 2002; Metcalfe et al. 2003). These examples cover only a small number of the possible applications of shape memory technology, which shows potential in numerous other applications. Commonly, SMPs are thermo-responsive materials that can convert thermal energy directly into mechanical work. Although this chapter focuses on thermal activation, interesting developments that involve the use of activation stimuli such as magnetic (Mirfakhrai, Madden, and Baughman 2007; Wang, Hu et al. 2007) or electric field (Lu and Evans 2002), irradiation (Ahir and Terentjev 2005; Lendlein, Jiang et al. 2005; Vaia 2005), and pH change/ionization (Harris, Bastiaansen, and Broer 2007; Mirfakhrai, Madden, and Baughman 2007; Rousseau 2004) have been reported.

Polymeric materials are intrinsically capable of a shape memory effect, although the responsible mechanisms differ dramatically from those of metal alloys. Compared with shape memory alloys (SMAs), SMPs achieve temporary shape deformation and recovery through a variety of physical means, and the underlying very large extensibility is derived from the intrinsic elasticity of polymeric networks. They offer the advantages of high elastic deformation (up to 400%), low cost, low density, and potential biocompatibility (the quality of not having toxic or injurious effects on biological systems) and biodegradability (the ability of degradation by living organisms). They also have a broad range of application temperatures that can be tailored, tunable stiffness, and are easily processed. SMPs also possess applications distinct from metal alloys due to their intrinsic differences in mechanical, viscoelastic, and optical properties.

The first SMP, a poly(norbornene)-based polymer, was reported by CDF Chimie Company, France, in 1984 and was made commercially available in the same year by Nippon Zeon Company of Japan (Liang, Rogers, and Malafeew 1997) under the trade name of Norsorex. But its applications have been limited by its processibility. Poly(urethane)-based SMPs were developed by Mitsubishi Heavy Industries in 1988 (Shirai and Hayashi 1988). The advantage of the poly(urethane) SMPs is the flexibility that the poly(urethane) chemistry provides in designing materials with a range of phase transition temperature. In addition, these poly(urethane)s are thermoplastic polymers that provide a significant improvement in processibility (Wei, Sandstrom, and Miyazaki 1998).

The basis of the thermally induced shape memory effect in polymeric materials is a large difference in Young's modulus of a material below and above a phase transition temperature. Figure 8.6 shows the Young's modulus of a polymeric material with phase transition temperatures—glass transition temperature (T_g) and crystalline melting temperature (T_m). The graph shows that the material is glassy hard below T_g,

---

[*] Please refer to: http://www.esato.com/archive/t.php/t-126804,1.html; http://gizmodo.com/190504/nokia-creates-self+destructing-cellphone. Nokia Research Center, together with a student group from Helsinki University of Technology, the Finnish School of Watchmaking, and the University of Art and Design, Helsinki, have developed a process for heat disassembly of portable devices. The idea is to disassemble a mobile phone by a heat-activated mechanism without any contact. By using a centralized heat source like laser heating, the shape memory alloy (SMA) actuator is activated, and the mobile phone covers are opened. The battery, display, printed wiring board (PWB), and mechanical parts are separated and can then be recycled in their material-specific recycling processes..

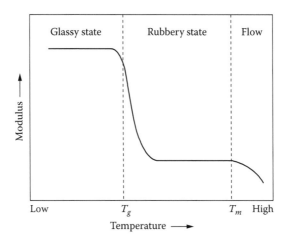

FIGURE 8.6 Modulus if a polymer had both amorphous and crystalline phases.

and becomes rubbery above T_g, but flows above T_m. Usually polymeric materials can be divided into two classes: one is amorphous polymer and the other is semicrystalline polymer. The polymer on the graph is semicrystalline polymer with very low degree of crystallinity. For an amorphous polymer, the modulus of the rubbery state usually shows slight decrease with increasing temperature. The T_g is defined as a temperature above which the molecular motion starts and the chain becomes flexible, and below which the molecular motion is frozen. When the polymer is semicrystalline with substantial degree of crystallinity, the rubbery state does not exist, and becomes flexible only above T_m. For a semicrystalline polymer, therefore, the T_m can replace the T_g in the modulus-temperature graph.

Figure 8.7 is the schematic representation of an amorphous polymeric molecular chain above T_g. Due to Micro-Brownian motion of chain segments, the chain becomes very flexible and shows random coil shape. The entropy of this random coil is very high and this shape becomes a thermodynamically stable shape at temperature higher than T_g. When the chain is extended by the stress, the entropy will be decreased. In the absence of any outside stress the molecular chain tends to maximize its entropy, so it will be recoiled again. This kind of behavior is called *entropy spring*.

Figure 8.8 is the schematic representation of the molecular motion of a rubber that consists of very flexible chain of very low T_g (such as below −50°C) and crosslinking points (or crosslinks). Flexible chains are tied by crosslinks, and a three-dimensional network is maintained. The rubber can expand several times under stress and return to its original shape and length almost immediately when the stress has been

FIGURE 8.7 Random coiled flexible polymeric molecular chain above its glass transition temperature.

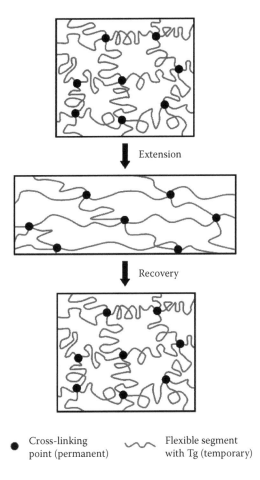

Extension

Recovery

● Cross-linking
point (permanent)

〜〜 Flexible segment
with Tg (temporary)

FIGURE 8.8 Schematic representation of polymer chains in a crosslinked rubber with extension and recovery.

removed. The shape memory effect of the SMP works due to the movement of a molecular chain, which is the result of the Micro-Brownian motion when heated above T_g (Kobayashi and Hayashi 1992; Shirai and Hayashi 1988). If the rubber is extended far more than the extension limit, then some chains between crosslinks will be broken, and the shape recovery will not be perfect. And this extension limit is usually very large compared to that of SMA. At ambient temperature rubber shows an elastic property, but elasticity is lost at temperature below its glass transition temperature (T_g). Consequently, the elongated shape is fixed and cannot return to its original shape below its T_g. At temperature above T_g, the polymer achieves a rubbery elastic state again.

Throughout the deformation and recovery process, the existence of crosslink is essential to maintain the network structure and remember the original shape of a polymer. Due to the crosslinks, relative connectivities of individual flexible chain are maintained, and the flexible chains become the only deformable part in the network.

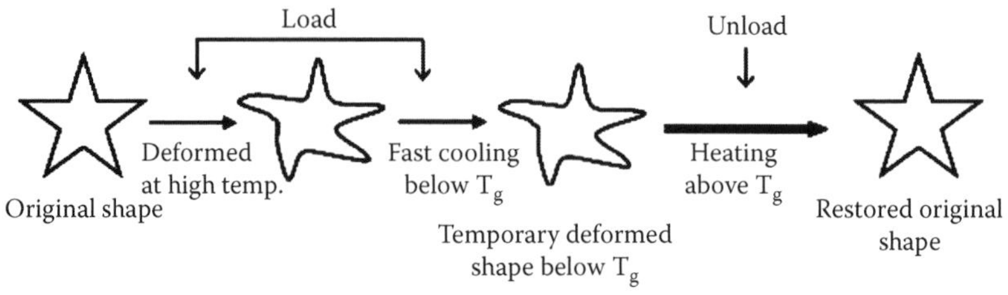

FIGURE 8.9   Thermo-mechanical cycle of an SMP.

In Figure 8.8, the network structure is maintained by the chemical crosslinking, but in other polymeric materials, network structure can be maintained by some different forms of crosslinking such as physical crosslinking.

A thermo-mechanical cycle of an SMP is shown in Figure 8.9. The cycle consists of four stages, marked as 1 through 4 in the Figure 8.9. In stage 1, SMPs can be deformed into a different shape quite easily when a stress is applied at a temperature above the transition temperature. In stage 2, deformed shape is maintained by applying the stress, and at the same time, the polymer is cooled to a temperature below the transition temperature and held at that temperature for a period of time. In stage 3, the sample is completely cooled. The force is no longer needed and the sample stays in the deformed shape. This is a temporary shape. At that state the polymer is fairly rigid and hard. In stage 4, the temporary state can be removed by heating the polymer above the transition temperature. During the transition the material goes from its temporary shape to its original memory shape. This is a very elastic and flexible state. This cycle of programming and recovery can be repeated several times with different temporary shapes in subsequent cycles. Therefore, in order to have a suitable shape memory effect, it is required to have a sharp transition from rigid state to rubbery elastic state, a long relaxation time, and a high ratio of glassy modulus to rubbery modulus.

A semicrystalline polymeric material has the crystalline melting transition temperature and goes from a solid state to a liquid state by melting. Concurrently, below and above the phase transition, a polymer goes from a high to low modulus state and the molecular chains become flexible random coils. That is, a crosslinked semicrystalline polymer can be used as an SMP. For crosslinked networks, therefore, the transition temperature (T_{trans}) can be the T_g or melting temperature (T_m). For both types of materials, the shape memory effect of the SMP works due to the movement of a molecular chain, which is the result of the Micro-Brownian motion that occurs by heating above T_{trans}.

Most of the shape memory effects are based on the existence of crosslinks that helps the polymeric chains go back and forth from the flexible random coils to the deformed rigid chains, while maintaining relative connectivities of individual flexible chains to remember the material's original shape. Therefore, as shown in Figure 8.8, the existence of crosslinks or net points is very important, and determines the permanent shape of the polymer network. The crosslink can be either a chemical crosslink (covalent bond) or a physical crosslink (intermolecular interaction), and this influences the morphology and thermo-mechanical property of the SMPs. In an SMP having chemical crosslinking, the volume or the size of a chemical crosslink is so small that the morphological phase of the SMP is practically considered as a single phase of flexible polymer chains at temperature higher than the transition temperature. When the crosslink is physical crosslink, however, the volume of the crosslink has to be substantial so that the SMP has two-phase morphology; one is the permanent phase (or hard segment), which acts as the physical crosslink, and the other is a reversible or switching phase (or soft segment).

In a physically crosslinked polymer system, there are many factors that can influence the properties of the SMP, such as the chemical structure, composition, and sequence-length distribution of the hard and soft segments in block copolymer, and overall molecular weight and its distribution. SMP can exhibit various thermally induced shape memory effects under control of these factors. For most physically crosslinked SMPs, the integrity of the crosslink is not perfect. Creep and some irreversible deformation can be observed during *programming* by extensive forces. The nature of the soft segment, on the other hand, determines the transition temperature of the reversible phase and can be of amorphous switching segment ($T_{trans} = T_g$) or crystalline switching segment ($T_{trans} = T_m$).

Consequently, SMPs can be classified into four categories depending on the nature of the crosslinked structure as well as the origin of their transition temperature (Liu, Qin, and Mather 2007; Rousseau 2004, 2008). They can be (1) chemically crosslinked amorphous thermosets, (2) chemically crosslinked semicrystalline rubbers, (3) physically crosslinked amorphous thermoplastics, and (4) physically crosslinked semicrystalline block copolymers. This classification also differentiates the mechanisms of shape fixity/recovery as well as of mechanical response of an SMP. Biodegradable SMP and SMP composite can also be classified into the four categories, but due to their importance in their specific applications, they will be treated separately following the four categories.

8.3.1 CHEMICALLY CROSSLINKED SHAPE MEMORY POLYMERS

SMPs based on chemical crosslinking have better shape memory performances compared to the physically crosslinked systems in terms of irreversible deformation due to creep during *programming.* Two synthetic methods are generally used for the preparation of the chemically (covalently) crosslinked SMPs (Figure 8.10).

First, it can be synthesized by chain addition copolymerization of vinyl monomer with di- or multifunctional monomer as shown in Figure 8.10a, or by step polymerization of difunctional monomer in the presence of multifunctional crosslinkers. In this case, the properties of the SMPs such as the thermal, chemical, and mechanical properties, are determined by the properties of the used monomers, their functionality, and the content of the crosslinkers. Copolymers made from various monomers or crosslinkers, which exhibit a shape memory behavior, were reported (Kagami, Gong, and Osada 1996; Goethals, Reyntjens, and Lievens 1998; Reyntjens, Du Prez, and Goethals 1999; Li, Perrenoud, and Larock 2001; Li and Larock 2002; Li, Hasjim, and Larock 2003; Liu and Mather 2002; Mather and Liu 2003; Tong 2002). Chemically crosslinked poly(urethane) SMPs, with amorphous segments or

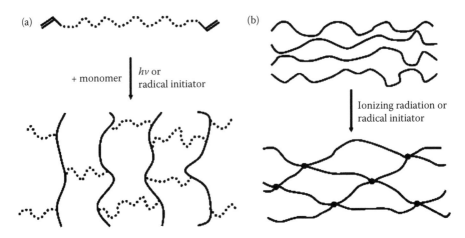

FIGURE 8.10   Two synthetic methods for the preparation of chemically crosslinked SMP.

semicrystalline segments, were prepared by using a various crosslinker (Buckley et al. 2006; Buckley, Prisacariu, and Caraculacu 2007; Xu, Shi, and Pang 2006; Chen, Zhu, and Gu 2002; Kim and Paik 1999; Lin and Chen 1999; Alteheld et al. 2005; Yang, Hu et al. 2006; Hu et al. 2005; Lee, Kim, and Kim 2004).

Second, chemically crosslinked SMPs can be obtained by the subsequent crosslinking of linear or branch polymers by a radical mechanism using ionizing radiation or radical initiators. The existence of radical-radical coupling makes it possible to create an intermolecular covalent bond between the polymer chains. The crosslinking density of the obtained polymers is exclusively dependent on the reaction condition and curing time. Shape memory effect of crosslinked polyethylene by means of ionizing radiation was investigated (Kumar and Pandya 1997; Chernous, Shil'ko, and Pleskachevskii 2004; Khonakdar et al. 2007; Li et al. 1998; Naga, Tsuchiya, and Toyota 2006). Crosslinked poly(vinyl chloride) by the thermal treatment under vacuum (Skakalova, Lukes, and Breza 1997), radiation crosslinked poly(ε-caprolactone) (Lendlein, Schmidt et al. 2005; Nagata and Kitazima 2006; Zhu et al. 2006; Zhu et al. 2003; Zhu et al. 2005), and radical initiator crosslinked poly[ethylene-*co*-(vinyl acetate)] (Li et al. 1999) were also investigated for the SMPs.

As mentioned previously, chemically crosslinked SMPs can be categorized into two classes by the difference in fixing mechanism according to the nature of switching segment/phase; one is chemically cross-linked amorphous thermosets, and the other is chemically crosslinked semicrystalline rubbers. Recent progress of chemically crosslinked SMPs will be introduced with discussion based on the two categories in the following section.

8.3.1.1 Chemically Crosslinked Amorphous Thermosets

A chemically crosslinked polymer forms a three-dimensional network structure and exhibits a sharp T_g at the temperature of interest and rubber elasticity above T_g. The permanent shape is set by the covalent bonds of the three-dimensional network during the crosslinking process. The temporary shape is commonly formed at T >

T_g by external stress and fixed by cooling below T_g. Usually, this type of material has attractive characteristics that include excellent shape fixity and recovery due to the high modulus below T_g and excellent rubber elasticity above T_g, tunable work capacity during shape recovery through the control of crosslinking density, and prevention of molecular slippage between chains due to strong chemical crosslinking. Nevertheless, these materials are difficult to reshape afterward once processed, because the permanent shape is set by chemical crosslinking. Good examples of this class are epoxy-based SMPs that have been reported in the literature to show fixing and recovery of 95%–100% (Gall, Kreiner et al. 2004; Liu, Gall et al. 2006; Liu et al. 2004). Such epoxy systems are commercially available thermoset systems with proprietary controlled molecular structures. Recently, Larock's group reported three-dimensional random copolymer networks of renewable nature oils, styrene, and divinylbenzene for shape memory behavior (Li, Perrenoud, and Larock 2001; Li and Larock 2002; Li, Hasjim, and Larock 2003). These networks show tunable glass transitions and rubbery properties upon variation of the monomer ratio. However, incomplete shape recovery occurred at around transition ranges due to broad glass transition spans and the coexistence of glassy and rubbery fragment, though complete shape fixing and recovery were observed at high temperature (Figure 8.11).

Some chemically crosslinked amorphous polyurethanes using various methods have also been investigated. Buckley, Prisacariu, and Caraculacu (2007) reported a novel thermosetting polyurethane using 1,1,1-trimethylol propane as a crosslinker (Figure 8.12). In this case, introduction of the covalent linkage results in the improvement in creep and increase in recovery temperature. Xu, Shi, and Pang (2006) synthesized hybrid polyurethane crosslinked with Si-O-Si linkages through hydrolysis and condensation of ethoxy silane groups. The Si-O-Si linkage acts not only as the net points but also as inorganic fillers for reinforcement. Chen, Zhu, and Gu (2002) and Kim and Paik (1999) prepared thermosetting polyurethanes having both shape

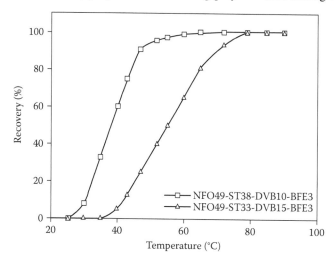

FIGURE 8.11 The shape recovery rates of fish oil ethyl ester (NFO), styrene, and divinylbenzene polymers as a function of temperature.

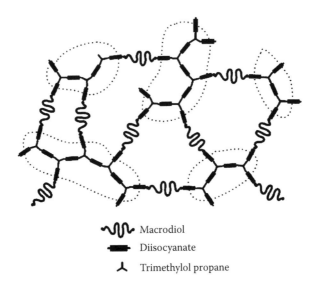

    ~\Jℓ~ Macrodiol

    ■■■ Diisocyanate

    ⅄ Trimethylol propane

FIGURE 8.12 Schematic diagram of the structure of the triol-crosslinked polyurethanes. Dashed lines indicate diisocyanate-rich regions of reduced mobility.

memory and hydrogel properties, by introducing hydrophilic poly(ethylene glycol) group into the crosslinked poly(urethane) backbone. Besides, crosslinked ester-type polyurethanes were reported by some workers (Lin and Chen 1999; Alteheld et al. 2005).

Some physically crosslinked polymers, with T_g above room temperature and with ultra-high molecular weight, exhibit characteristics like the thermosetting SMPs due to their lack of flow above T_g by high entanglement density (>25) and good shape fixity by vitrification. These materials have a three-dimensional network structure that gives excellent elasticity above T_g, but makes thermal processing difficult. For example, poly(norbornene) (PN) (Mather, Jeon, and Haddad 2000; Nagai, Ueda, and Isomura 1994) with $T_g \sim 40°C$ and high molecular weight poly(methyl methacrylate) (PMMA) (Yang, Zhang, and Li 1997; Beloshenko et al. 2002) with $T_g \sim 105°C$ show quite complete shape fixing when vitrified and demonstrate fast and complete shape recovery due to the sharp glass transition temperature. They form a three-dimensional network, evidenced by a flat rubbery plateau measured rheologically.

In addition to the examples given above, other polymers are investigated to be SMPs based on the same mechanism, such as dehydrochlorinated crosslinked poly(vinyl chloride) (Skakalova, Lukes, and Breza 1997), optically transparent poly(methyl methacrylate-*co*-butyl methacrylate) (Liu and Mather 2002; Mather and Liu 2003), styrene copolymer (Tong 2002), HDI-HPED-TEA network (Wilson et al. 2005), and poly[(methyl methacrylate)-*co*-(*N*-vinyl-2-pyrrolidine)]-poly (ethylene glycol) semi-IPNs (Liu, Guan et al. 2006).

8.3.1.2 Chemically Crosslinked Semicrystalline Rubbers

Similar to an amorphous thermoset SMP having glass transition as a transition temperature, temporary shape of a semicrystalline rubber can be fixed when the sample is deformed above T_m of the crystalline regions and subsequently cooled below their crystallization temperature. Similar to the glassy materials, the permanent shapes are established by chemical crosslinking and cannot be reshaped after processing. This class of materials is usually more stable below transition temperature and their shape fixity and recovery properties depend on the degree of crystallinity. Generally, these materials include bulk polymers, such as semi-crystalline rubbers, liquid crystalline elastomers (LCEs), and hydrogels with phase-separated crystalline microdomains.

Generally, chemically crosslinked semicrystalline SMPs are synthesized by subsequent crosslinking of a linear or a branched polymer (Kumar and Pandya 1997; Chernous, Shil'ko, and Pleskachevskii 2004; Khonakdar et al. 2007; Li et al. 1998; Naga, Tsuchiya, and Toyota 2006; Skakalova, Lukes, and Breza 1997; Li et al. 1999). The influence of the degree of crosslinking on the gel content and heat shrink properties of γ-ray crosslinked polyethylene was described for low density polyethylene (LDPE) and high density polyethylene (HDPE) (Kumar and Pandya 1997). Also, crosslinked polyethylene systems (Khonakdar et al. 2007; Li et al. 1998) were used as heat shrinkable materials with fixing and recovery of up to 96% and 94%, respectively. Crosslinked ethylene-vinyl acetate rubber is produced by treating the radical initiator dicumyl peroxide (DCP) in a thermally induced crosslinking process (Li et al. 1999). Materials with different crosslinking density are obtained depending on the initiator concentration, crosslinking temperature, and the curing time, hence strain recovery in the range of 30%–95% was observed (Figure 8.13). The shape memory effect has also been investigated for radiation crosslinked poly(caprolactone)s (Lendlein, Schmidt et al. 2005; Nagata and Kitazima 2006; Zhu et al. 2006; Zhu et al. 2003). The cross-linked poly(caprolactone) has a higher tensile modulus and heat resistance, and exhibits excellent shape memory properties. In order to improve the crosslinking efficiency, the blends of poly(caprolactone) and polyfunctional acrylate monomers have also been reported (Zhu et al. 2005). Mather et al. (Liu et al. 2002) reported a chemically crosslinked, semicrystalline poly(cyclooctene) with a trans

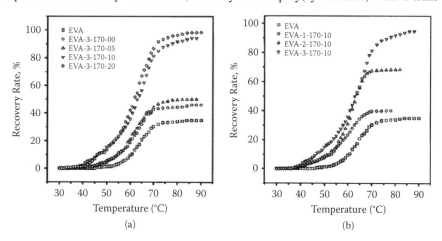

FIGURE 8.13 Strain recovery curves for EVA copolymers cured at 170°C; (a) different times, (b) different DCP contents.

content of 80%, a T_g of -70°C, a T_m of 58°C, and much better thermal stability. Especially, in the case of poly(cyclooctene) containing either 2.5 or 5 wt percent peroxide, complete shape recovery from a curvature to zero curvature occurred within 0.7 s at 70°C.

Besides, chemically crosslinked semicrystalline SMPs can be synthesized via copolymerization of monofunctional monomer with low molecular weight or oligomeric crosslinker. Covalently crosslinked copolymers were prepared by copolymerization of stearyl acrylate, methacrylate, and *N,N'*-methylenebisacrylamide as a crosslinker (Kagami, Gong, and Osada 1996). The thermal transition triggering the shape memory behavior is the melting temperature of the crystalline domain formed by the stearyl side chain that can be controlled by changing monomer composition, which enables one to adjust shape memory effect at a desired temperature. Goethals's group produced multi-block copolymer networks by radical copolymerization of poly(octadecyl vinyl ether) diacrylates or –dimethacrylates with butyl acrylate (Goethals, Reyntjens, and Lievens 1998; Reyntjens, Du Prez, and Goethals 1999) (Figure 8.14). Segmented networks containing poly(octadecyl vinyl ether) and poly(butyl acrylate) segments show a high degree of phase separation, which allows the octadecyl vinyl ether units to crystallize, and hence exhibit shape memory properties. However, the strain recovery rate was not quantified more precisely with respect to the composition of the monomers. Crosslinked poly(urethane)s were made by using excess diisocyanate or by using a crosslinker like glycerin, trimethylol

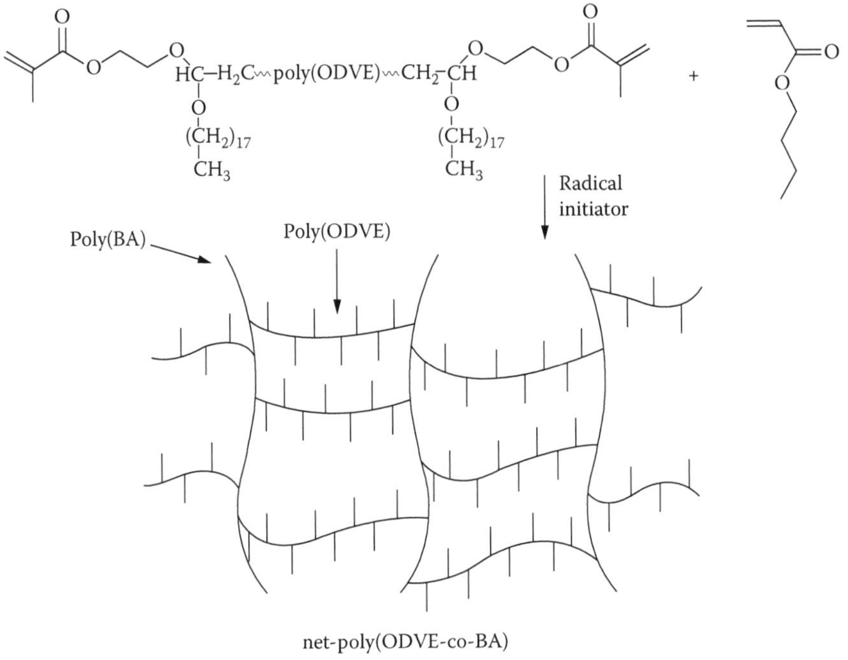

net-poly(ODVE-co-BA)

FIGURE 8.14 The network formation of multi-block copolymer of poly(octadecyl vinyl ether) and poly(butyl acrylate).

propane (Yang, Hu et al. 2006; Hu et al. 2005; Lee, Kim, and Kim 2004). In the case of poly(urethane) with crystalline soft segments, chemical crosslinking decreases the crystallization of the soft segment and improves the mechanical properties of the resulting material. Liquid crystalline polymers can also be used for shape memory application. Rousseau and Mather (2003) synthesized a smectic liquid crystalline elastomer (LCE) as an SMP and shape memory effect was characterized using a shape memory cycle test. This LCE features a sharp transition close to the body temperature, a low modulus comparable to muscle, and optical clarity.

Although chemically cross-linked semicrystalline rubbers can be tailored to optimize performance and reach almost complete shape fixity and recovery, the modulus in the fixed state is relatively lower than that for amorphous thermosets because the temporary shape is fixed through crystallization. Furthermore, large thermal hysteresis between melting and crystallization may lead the cooling temperature to lower temperature than T_s in order to allow full crystallization for good shape fixity, and this potentially extends the shape memory cycle time (Rousseau 2008).

8.3.2 Physically Crosslinked Shape Memory Polymers

Generally, SMPs have two separated phases (the thermally reversible phase and the frozen phase) (Lendlein and Kelch 2002; Lin and Chen 1998b; Kim et al. 2000; Kim and Lee 1996; Kim et al. 1998; Jeong, Lee, and Kim 2000). The thermally reversible phase shows a lower phase transition temperature (T_{trans}), which can either be a glass transition (T_g) or a melting transition (T_m) temperature. It serves as a molecular switch and enables the fixation of a temporary shape. The frozen phase shows a higher thermal transition temperature (T_p) and works as the physical crosslink that is responsible for remembering the permanent shape (Figure 8.15). When SMPs are heated to the temperature $T_{trans} < T < T_p$, large deformations can easily be developed for the polymers with low modulus. If the SMPs are cooled at a temperature below T_{trans}, such deformations can be consequently fixed due to the dramatic increase in the modulus. The original shape can be recovered under the entropic elasticity by being reheated up to a temperature above T_{trans} (Lendlein and Kelch 2002).

The process sequence in Figure 8.15 can be as follows:

1. Deformation (heating)—deforming the SMPs in the temperature range $T_{trans} < T < T_p$.
2. Fixation (cooling)—fixing the temporary shape of the SMPs by cooling to below T_{trans} and releasing the constraint.
3. Recovery (reheating)—heating the SMPs above T_{trans} to recover the original shape.

It is well known that the chemical crosslinking of polymers severely lowers the extent of molecular motion and results in poor processibility of the polymers. In order to improve the processing conditions and manipulate the structure of shape memory polymers, researchers have tried to study shape memory polymers with physical crosslinking instead of chemical crosslinking. Also, the advantage of using physically crosslinked segmented copolymers as shape memory materials is

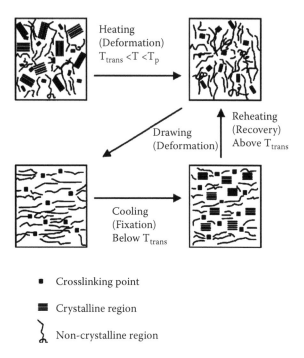

FIGURE 8.15   Shape memory effect of physically crosslinked polymers.

the possibility of controlling the critical recovery temperature to meet the needs of different applications, so the shape memory polymers are investigated for the purpose of improving processing conditions of their preparation and widening the list of polymers for shape memory applications. But physically crosslinked SMPs generally exhibit slightly lower shape memory performances compared to chemically crosslinked glassy thermosets or chemically crosslinked semicrystalline rubber SMPs, especially in terms of shape recovery and retention. The reason is mainly explained by a decreased physical crosslink integrity caused by mechanical deformation (Jeong, Lee, and Kim 2000). However, physically crosslinked SMPs exhibit a relatively high modulus below T_{trans}, comparable to that of chemically crosslinked glassy thermoset SMPs.

The majority of physically crosslinked SMPs is represented by segmented shape memory polyurethanes (SMPUs), which are thermoplastic block copolymers having peculiar mechanical properties and thermal-responsive shape memory effects, which are due to the presence of soft and hard segments (Figure 8.16)

Hard segments bind each other through interactions such as hydrogen bonding, dipole–dipole interaction, and van der Waals forces, and serve as physical crosslinking points and are responsible for recovering the original shape after deformations (Cho, Jung, Chung et al. 2004). Soft segments absorb external stress during elongation or compression and decide the phase-transition temperature (T_g or T_m). Completion of shape retention and recovery by the balanced composition of hard and soft segments results in reversible shape conversion around the phase-transition temperature.

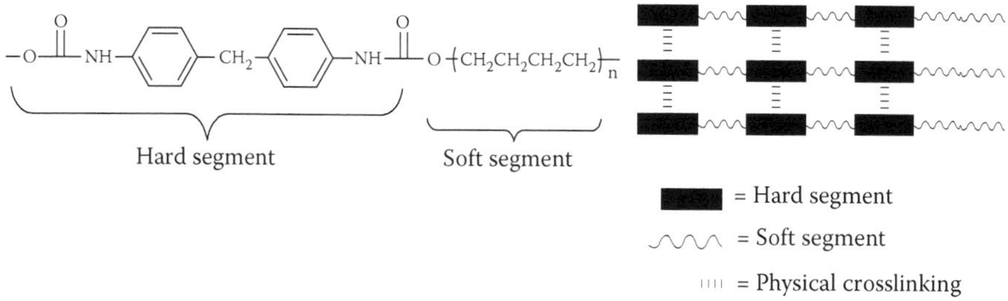

= Hard segment
= Soft segment
= Physical crosslinking

FIGURE 8.16   Segmented shape memory polyurethanes (SMPs) composed of hard and soft segments.

8.3.2.1 Physically Crosslinked Amorphous Thermoplastics

The shape memory behavior is attributed to the phase transition temperature (T_g) of the soft segment (elastic amorphous matrix, which has low T_g) regions for physically crosslinked amorphous thermoplastics. A deformed shape at $T > T_g$ is maintained by cooling below the glass transition. The incorporation of ionic or mesogenic moieties into the hard segment phase can influence the mechanical and shape-memory properties through the introduction of additional intermolecular interactions by possibly enhancing the degree of phase separation and integrity of the hard phase. A representative example of the physically crosslinked amorphous thermoplastics is polyether-based polyurethane, which has amorphous soft polyether segments and aromatic urethane hard segments. The material was synthesized by using the prepolymer method, by polymerizing alternatively hard segment (diphenylmethane diisocyanate) and soft segment [poly(tetramethylene glycol)] (Figure 8.17). And various interactions among the segments led to the domain formation (Kim and Lee 1996; Li et al. 1997; Li et al. 1996) (Figure 8.18).

The permanent shape of the network is provided by physical crosslinking of the hard segments through molecular interactions such as hydrogen bonding, dipole–dipole interactions, or van der Waals forces. In order to have effective shape memory property, the hard segments should maintain the shape through inter- or intra-polymeric chain attractions.

But Yang et al. (Yang, Huang et al. 2006) carried out systematic research on the effects of moisture on the T_g and thermo-mechanical properties of a polyether-based polyurethane. The results showed that the hydrogen bonding between N-H and C=O groups is weakened by the absorbed water, which resulted in considerable deterioration of shape memory properties (Lim, Britt, and Tung 1999) (Figure 8.19).

The T_g of thermoplastic polyurethane materials could be controlled in a wide range from –30°C to 100°C by using different kinds of urethane ingredients (diisocyanate, polyol, and chain extenders) and by adjusting their molar ratios (Lee et al. 2001; Yang, Huang et al. 2006; Mondal and Hu 2007; Takahashi, Hayashi, and Hayashi 1996; Kim et al. 1999; Park et al. 2004; Lin and Chen 1998a, 1998b; Jeong, Lee, and Kim 2000; Chun, Cho, and Chung 2007; Wang and Yuen 2006; Yang et al. 2003; Cho, Jung, Chun et al. 2004; Cho, Jung, Chung et al. 2004). In the segmented polyurethanes, the hard segment acts as the physical crosslink, and so hard segment

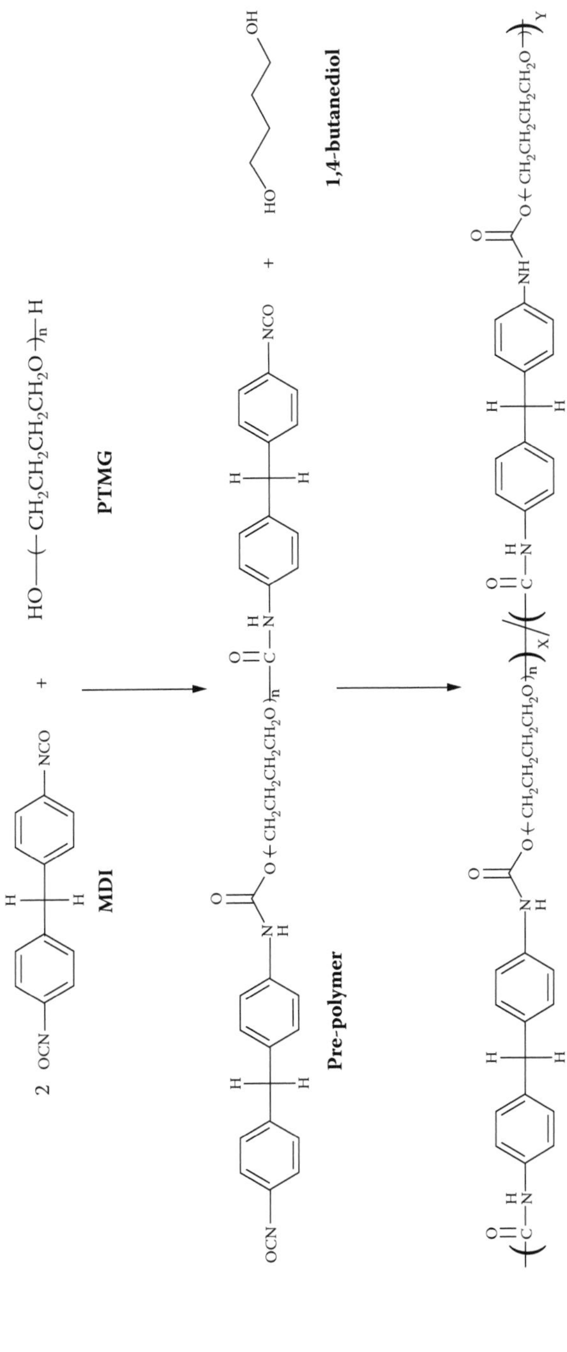

FIGURE 8.17 Pre-polymer method for the synthesis of thermoplastic polyurethanes.

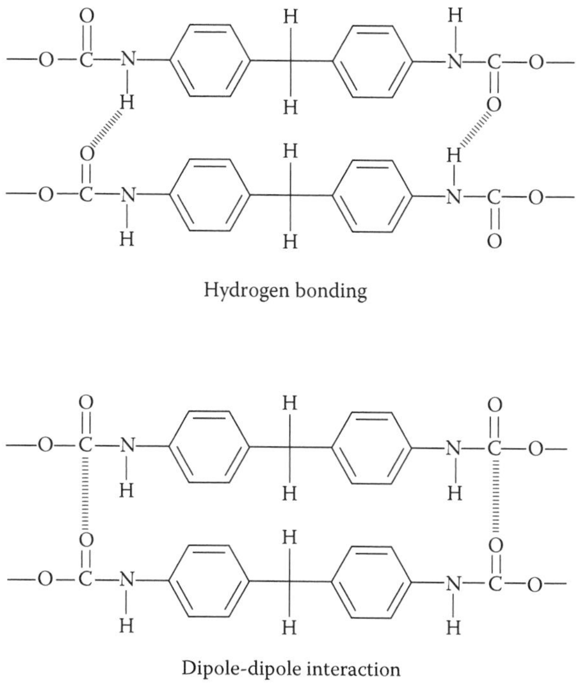

Hydrogen bonding

Dipole-dipole interaction

FIGURE 8.18 Interactions among polymeric chains.

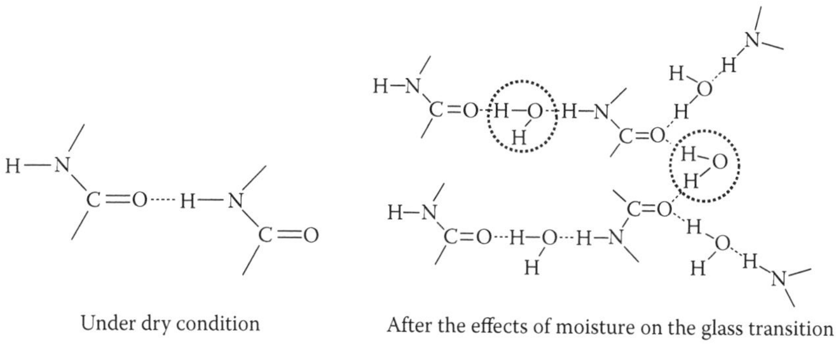

Under dry condition   After the effects of moisture on the glass transition

FIGURE 8.19 Effects of water on the hydrogen bonding in polyurethane SMPs.

concentration should be sufficiently high (>20 wt %) for shape memory effect (Lee et al. 2001). Optimum shape memory properties (harmonic composition of shape retention and recovery) were achieved at 35–40 wt % of hard segment concentration (Figure 8.20).

Hard segment works as a pivoting point for shape recovery and soft segment could mainly absorb external stress applied to the polymer by unfolding and extending their molecular chains (Hu, Ji, and Wong 2005). If the stress exceeds and breaks

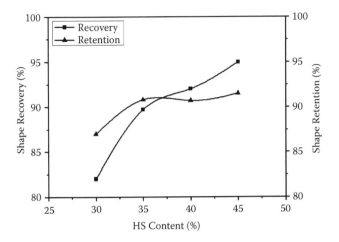

FIGURE 8.20 Shape recovery and retention curves (PU-based SMPs) plotted against hard segment concentration.

the interactions among hard segments, shape memory will be lost and original shape cannot be recovered. Therefore, precise control of composition and structure of hard and soft segments is very important to satisfy the conditions required for various applications. For this control, mechanical properties of polyurethane-based shape memory polymer coatings were investigated (Cho, Jung, Chun et al. 2004) and the studies show that the mechanical property was influenced by the polyurethane hard segment content as shown in the Table 8.1.

But especially for the water vapor permeability, only a slight change was observed with the control of polyurethane hard-segment content in the property of polyure-thane-coated fabrics. Except for water vapor permeability of polyurethane films, the combination of such structure-controlling factors can also be used to tailor the glass-transition temperature of shape-memory polyurethane for specific uses, lead-ing to application of polyurethane in other fields such as medicine, industry, sports, and textiles. Blending of polyurethane with other polymers, as a means to satisfy the

TABLE 8.1
Mechanical and Thermo-Mechanical Data of PU-Based Block Copolymer

Hard Segment Content (%)	Breaking Stress (MPa)	Breaking Elongation (%)	Modulus (MPa)	Shape Retention	Shape Recovery
50	7.2	52	33	—	—
40	5.3	120	14	92	95
30	3.9	165	13	90	83

Source: Cho, J. W., Y. C. Jung, B. C. Chun, and Y. C. Chung. 2004. Water vapor permeability and mechanical properties of fabrics coated with shape-memory polyurethane. *Journal of Applied Polymer Science* 92 (5): 2812–16. Reprinted with permission of John Wiley & Sons, Inc.

demands of improved mechanical properties, has been studied by changing the blend compositions. But literature about shape memory effect of polymer blends is limited (Jeong, Ahn, and Kim 2001; Jeong et al. 2001; Kusy and Whitley 1994) probably due to the versatile polyurethane chemistry, which offers a wide range of tailor-made properties by employing various materials.

8.3.2.2 Physically Crosslinked Semicrystalline Block Copolymer

The molecular structures are very similar to physically crosslinked amorphous SMPs described above, except for the fact that the soft domain will crystallize and, instead of T_g, their T_m values function as shape-memory transition temperatures and the secondary shapes are fixed by crystallization of the soft domains.

Parameters describing the recovery temperature, ability, and speed were used to study the influence of structure and processing conditions on the shape memory behavior of the polymers. It was found that (1) the high crystallinity of the soft segment regions at room temperature and (2) the formation of stable hard segment domains acting as physical crosslink in the temperature range above the melting temperature (T_m) of the soft segment crystals are the two necessary conditions for a segmented copolymer with shape memory behavior. The final recovery rate and the recovery speed are mainly related to the stability of the physical crosslinks under external stress and are dependent on the hard segment content of the copolymers.

Conventionally, polyurethanes are multiblock copolymers consisting of alternating oligomeric sequences of hard and soft segments. The hard segments are responsible for physical crosslink by way of polar interaction, hydrogen bonding, or crystallization, with such crosslink being able to withstand moderately high temperatures without being destroyed. On the other hand, crystallizable soft segments are in charge of the thermally reversible phase and the crystallization of these soft segments controls the secondary shape. Polyurethanes feature the advantage of easy tunability of room-temperature stiffness, transition temperature, biocompatibility, and mechanical strength by controlling their compositions. Polyurethanes can easily be foamed, as with the foamed shape-memory material *cold hibernated elastic memory (CHEM)*, which was successfully manufactured and brought to application (Sokolowski and Chmielewski 1999, reproduced by permission of The Royal Society of Chemistry).

The most commonly reported systems are thermoplastic-segmented polyurethane based on polycaprolactone diol (PCL) as a soft segment for crystallization, methylene diisocyanate (MDI) as a hard segment, and butandiol (BD) as the chain extender (Cao and Jana 2007; Li et al. 1997; Li et al. 2000). Figure 8.21 shows the synthetic scheme of PCL-diol.

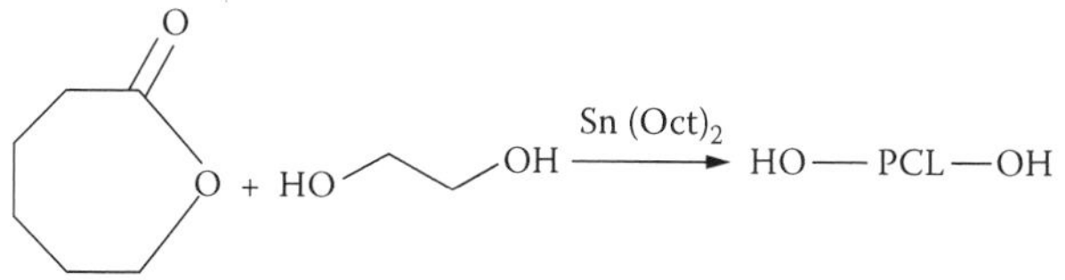

FIGURE 8.21 Reaction for synthesis of PCL-diol.

This shape-memory effect can be controlled via the molecular weight (block length) (Lin and Chen 1998a, 1998b; Kim et al. 2000) of the soft segment, the molar ratio of copolymer composition (the hard and soft segments), crystallization of segments, and the polymerization process. For SMPs with semicrystalline switching segment, PCL segments undergo micro-phase separation. The recovery temperature can vary from 40°C to 60°C depending on the molecular weight of PCL diol segments and soft/hard segment composition. The relation between the shape memory effect and molecular structure has been studied and it was concluded that high crystallinity of the soft segment region at room temperature is a necessary prerequisite for a good shape-memory behavior (Lendlein, Kelch, and Kratz 2006; Jeong et al. 2000).

For the dependence of PCL crystallinity on the molecular weight of soft segments, no crystallinity was observed for samples with soft segments of low molecular weights between 1600 and 2000 g/mole. The crystallinity of PCL segments increases abruptly after a critical molecular weight value, then it approaches a constant. This critical molecular weight is in the range of 2000–3000 g/mole and slightly increases with increasing hard-segment length, and it might be the lower limit for molecular weight of PCL segments that can be used for preparing segmented polyurethanes having a shape-memory effect (Figure 8.22).

When PCL-diol has a number average molecular weight less than 2000 g/mole, the crystallization of PCL segments is hindered by incorporating them into the multiblock copolymers. It is clear that the molecular weight of the soft segments is the main factor determining the crystallizability. For PCL segments of higher molecular weight, the influence of copolymerization on the crystallization behavior was found in the lower crystallization temperature (T_c) and the decrease of crystallinity. If the molecular weight increases, the influence for the property decreases. The crystallization ability of PCL segments in segmented polyurethanes is significantly depressed due to the connection with the hard segments. The optimum molecular weight in

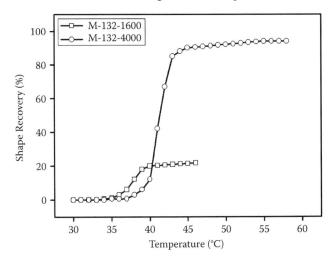

FIGURE 8.22 Thermal recovery curves of polyurethanes with PCL segments of molecular weight of 1600 and 4000.

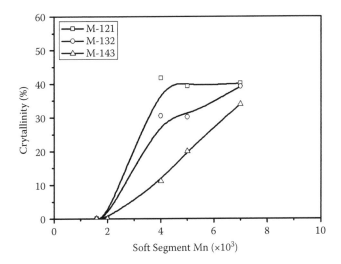

FIGURE 8.23 Dependence of PCL crystallinity of polyurethanes on molecular weight of soft segments.

terms of shape memory properties was found to be 4000–6000 g/mole (Jeong et al. 2000; Kim and Lee 1996) (Figure 8.23).

Shape memory effects have also been reported for shape memory polymers with other crystalline phases. Typical examples are trans-poly(isoprene-based polyurethane [PU]) (Ni and Sun 2006; Sun and Ni 2004) and poly(ethylene oxide) (PEO)/poly(ethylene terepthalate) (PET) (Luo et al. 1997; Wang et al. 1997; Wang and Zhang 1999). The melting point of PEO varies from 45°C to 65°C, depending on the molecular weight. However, the melting point of PEO of same molecular weight (in the block copolymer) decreases with increasing PET content due to the steric-hindrance for the crystallization process typically observed in block copolymers. Styrene–trans-butadiene–styrene (STBS) triblock copolymers feature shape memory behavior that is afforded by this mechanism (Ikematsu, Kishimoto, and Karaushi 1990) (Figure 8.24).

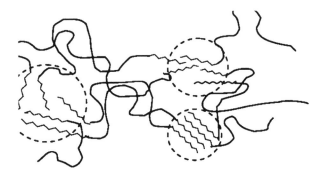

FIGURE 8.24 Representation of hard block aggregation in ABA triblock copolymer.

Also, researchers already have designed, synthesized, and fully characterized a series of novel triblock, multiblock polyurethane copolymers for shape memory applications using the polyhedral oligomeric silsesquioxane (POSS) hybrid monomer in the hard domains and various polyols, either amorphous or semicrystalline, as soft domains (Mather et al. 2004; Kelly and Mather 2005; Mather et al. 2006; Qin and Mather 2006). Studies on segmented block copolymers and graft copolymers shown above indicate that it is possible to use physical crosslink in shape memory polymers instead of chemical crosslink. However, physical crosslink is usually not as strong or as stable as chemical ones. So it is of great interest and importance to find out whether the unstable nature of the physical crosslink will influence the shape memory effect of these polymers and to what extent this influence will be.

8.3.3 Biodegradable SMP

Most shape memory polymers are not biodegradable regardless of whether they are crosslinked chemically or physically. For medical implant applications, additional properties are required such as biodegradability and the recovery temperature around the human body. If a medical item is biodegradable, it can stay in a human body for a desired period. It means that a secondary surgery is not required. Since Lendlein and Kelch reported shape memory polyurethane (PU) based on poly (ε-caprolactone) (PCL) and demonstrated its potential in medical applications (Lendlein and Kelch 2002), biodegradable shape memory polymers have become a considerable research interest. Most biodegradable shape memory polymers reported recently are based on PCL, polyglycolide (PGA), poly (L-lactide) (PLLA), or others that are well-known biodegradable polymers. Most biodegradable SMPs based on these polymers have good shape memory properties, but there are some drawbacks. They have low recovery stress or they have poor mechanical strength after shape recovery. For PCL-based polymers, by adjusting the molecular weight of PCL diol and the hard-to-soft ratio, the lowest recovery temperature can be adjusted to 37°C–42°C (Lendlein and Kelch 2002; Ping et al. 2005). Figure 8.25 shows one example of synthetic scheme for PCL-based polyurethane. PCL-based polymer has one disadvantage: poor mechanical strength after shape recovery. For PLA-based polymer, this disadvantage can be improved (Wang et al. 2006). But its glass transition temperature of around 60°C is too high to use in a human body. Wang et al. (Wang, Ping et al. 2007) prepared a series of biodegradable polyurethanes using the co-oligomer of lactide and ε-caprolactone as soft segments, 2,4-toluene diisocyanate (TDI) as hard segment, and 1,4-butanediol (BDO) as chain extender. This copolymer offers recovery temperature adjustment through PCL component and degradation rate adjustment through PLA component. Their T_gs are in the range of 28°C~53°C, influenced by the CL content in PCLA.

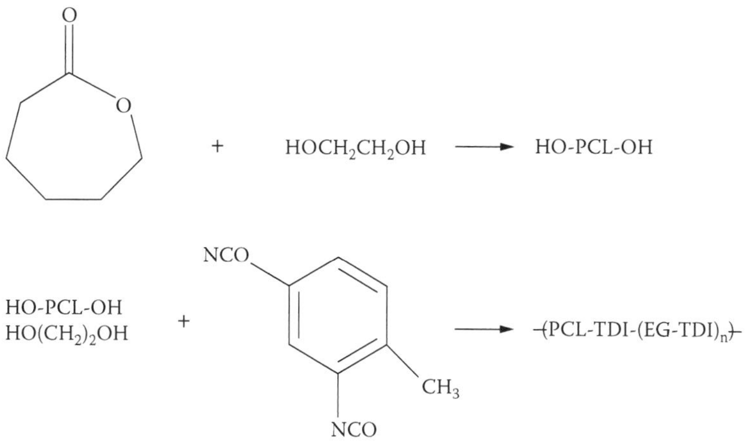

FIGURE 8.25 Synthetic scheme of a PCL-based polyurethane.

8.3.4 SMP COMPOSITE

SMPs have the advantages of light weight, low cost, and excellent processability. But pure SMPs have lower strength and stiffness than metallic shape memory alloys. And the recovery force is low due to low stiffness. SMPs, therefore, cannot be used in some applications that require high strength, high stiffness, high recovery force, and good resistance against creeping. To improve these features, many studies are reported for SMP composites. Incorporation of reinforcing fillers has been investigated to increase mechanical properties of SMPs. The nanoparticle of SiC was used as a filler to epoxy-based SMP composites to improve strength and recovery force (Gall et al. 2002; Gall, Dunn et al. 2004). Huang and Yang reported the carbon black filled polyurethane (PU)-based SMP composites in order to improve electric conductivity (Yang, Huang et al. 2006). Carbon nanotubes as fillers were also reported to improve mechanical and electrical conductive properties (Sahoo et al. 2005). The nanocomposites offer electrically induced actuation. It is important in many practical applications such as smart actuators for controlling micro vehicles. To obtain conducting SMPs, multi-walled carbon nanotubes were used after the surface was chemically modified in a mixed solvent of nitric acid and sulfuric acid, for improvement of interfacial bonding between polymers and conductive fillers. These reports show how to fabricate electro-active SMP nanocomposites and how to utilize the voltage triggering shape memory for application as an actuator.

8.3.5 APPLICATION OF SMP

As mentioned above, shape memory has many useful abilities. But it's only a small part of SMP. SMP has quite large capability such as moisture permeability, changes in specific volume, and refractive index. These factors make SMPs intelligent and adaptive materials. Some of the applications reviewed in the following three parts are medical devices (Sokolowski, Metcalfe, and Hayashi 2007), clothing materials, and aircraft technology.

8.3.5.1 Medical Devices

Shape memory polymer has high potential in its applicability to our life. The properties of SMPs are suitable to various products in microelectromechanical systems, for self-healing (e.g., stent) and health monitoring purposes, and in biomedical devices. Polyurethane-based shape memory polymers have been considered to be used in medical applications. They have two unique properties. One is a good biocompatibility when in contact with body fluids, as shown through standard cytotoxicity and mutagenicity tests. The other is the glass transition temperature, which can be close to the human's body temperature.

Hayashi et al. had investigated SMP for several different medical applications. The catheters made from SMP materials remain stiff externally for accurate manipulation by the physician, but become softer and more comfortable inside the human body (Hayashi 1993). Soft catheters cause fewer injuries than rigid catheters, and are easier to manipulate and conduct in tortuous vessels. SMPs are also considered to be used for orthopedic braces and splints that can be custom fitted to requirements. By heating the shape memory component above its T_g and deforming it, a desired requirement can be obtained and then sustained after cooling. In one of the first applications, a polyurethane-based SMP has been used for a spoon handle designed for the handicapped. It can be heated and deformed to the shape of individual hands and then its deformation is fixed at room temperature to be a comfortable and custom fit.

In order to be considered as medical devices, some specific properties of SMPs are important in addition to the mechanical and thermal properties. Properties such as moisture permeability, energy dissipation, and storage may be important for some applications like bandages or artificial skin. Because the loss tangent, tan δ, in the transition region of SMP is possible to be very close to that of human skin, these types of advances may create new opportunities to enable a bulky device to be implanted into the human body but maintain a natural and smooth feel.

Shape memory polymer can be substantially different from an initial temporary phase due to the ability to memorize a permanent shape. A bulky device that has a temporary shape like a string could potentially be introduced into the body that could go through a small laparoscopic hole at low temperature and then expanded on demand into a permanent shape at body temperature.

Recently, Wache et al. conducted a feasibility study and preliminary development on a polymer vascular stent with shape memory as a drug delivery system (Wache et al. 2003). Samples from thermoplastic polyurethane-based SMP were manufactured by injection molding and the field of applications of a polymer stent was showed in pre-trials. Presently, almost all commercially available stents are made of metallic materials. There are several designs of these minimally invasive implanted vascular stents for coronary applications among which are tubular mesh, slotted tubes, and coils. A common aftereffect of stent implantation is restenosis. A drug delivery system using the shape memory polymer leads to significant reduction of restenosis and thrombosis. An improved biological tolerance in general can be expected when utilizing biocompatible SMP materials.

Another application is a biodegradable shape memory polymer for wound closure (Lendlein and Langer 2002). A design of a smart surgical suture has been considered whose temporary shape is obtained by elongating the fiber with controlled stress. This suture is applied loosely in its temporary elongated shape. When the temperature is raised above T_g, the suture shrinks and tightens the knot, applying the optimum force. In this application, removal of the implant in follow-up surgery is also not necessary. The wound was loosely sutured at low temperature. And when the temperature was increased to the body temperature, the shape-memory effect was started and the suture tightened the knot (Lendlein and Langer 2002).

SMP materials could be used in a variety of different medical devices and diagnostic products as deployable elements of implants from vascular grafts to components of cardiac pacemakers and artificial hearts. Present memory metals such as Ni-Ti are being used as components of different devices and provide a means of inserting a thin, wire-like device, contained in a needle-like casing, through a small incision. This device can regain a more complex shape once the case is removed.

The concept, called cold hibernated elastic memory (CHEM), was developed by Sokolowki et al. as a new smart structure that is light, self-deployable, and low cost. Many CHEM medical applications are considered for vascular and coronary grafts, catheters, orthopedic braces and splints, medical prosthetics and implants, and others (Sokolowski 2005). This CHEM has also been considered for use in spacecraft. It represents the next generation self-deployable structure and is intended to be supplemental to space inflatable structure technology.

8.3.5.2   Clothing

Recently, one of the textile applications for shape memory polymer was a shirt designed by Corpo Nove, a fashion house in Italy (Clowes 2001). The shirt, with long sleeves, could be programmed so that the sleeves shorten as room temperature becomes hotter. The fabric can be rolled up, pleated, creased, and then returned to its former shape by applying heat. It also does not require ironing. It is woven from fibers of the shape-memory alloy nitinol, interspersed with nylon. When heated to a certain temperature, the shirt returns to its original shape.

Recently, interest in electrospinning has been increasing rapidly because the polymer fibers produced by this method have submicrometer diameters, which cause them to be of considerable value in a wide variety of applications such as filtration, reinforcements in composites, and biomedical devices. Havens, Snyder, and Tong (2005) reported that electrospun nonwovens of shape memory polyurethane block copolymers with hard segment concentrations of 40 and 50 wt % were successfully prepared by electrospinning processing. These have more than 80% shape recovery abilities.

Vili (2007) manufactured the SMP-based smart woven fabrics for a special design. A sample for three-dimensional surface effect is solely woven with SMP yarn formations in regular floats. When the sample is stimulated the whole sample contracts and enhances the design by creating more depth and a three-dimensional texture. These designs show additionally effective shape on the reverse side when a circle pattern is created. These features are very valuable in the design of clothing for interior applications.

One application of the SMP in the textile industry is to manufacture intelligent, waterproof, breathable fabrics. Its fabrics could restrict the loss of body warmth by stopping the transfer of vapor and heat, and at the same time, could transfer more heat and water vapor from inside clothing to outside than ordinary waterproof, breathable fabrics at high temperatures. Ding, Hu, and Tao (2004) indicated the relationship between water vapor permeability of shape memory polyurethane film and temperature. When the temperature rises from 10°C to 40°C, water vapor permeability increase significantly.

Mitsubishi Heavy Industries has produced an intelligent clothing called DiaPLEX. It takes advantage of Micro-Brownian motion (thermal vibration). Micro-Brownian motion occurs within the fabric membrane when the temperature rises above a predetermined activation point (Figure 8.26). The SMP laminate is placed between two layers of fabric, creating a membrane, which is simultaneously waterproof and breathable. When ambient temperature is below activation point molecular structure is rigid, so permeability is low and body heat is retained. When ambient temperature is above activation point Micro-Brownian motion creates gaps between molecules, increasing permeability so that moisture and body heat can escape

8.3.5.3 Aerospace Engineering

A recent move in aerospace and space technology is taking advantage of SMP. Developing and demonstrating morphing materials and technologies that are necessary to construct deployable morphing aircrafts and other innovative adaptive structures critical to the Air Force are taking place, for example, Air Force morphing structural applications, and folding wing aircraft with wings fabricated completely from the shape memory composites called Veritex and locally transitioned along discrete lines to enable folding of wings (Havens, Snyder, and Tong 2005) (Figure 8.27).

There are lots of products in our lives. Some items are useful for ergonomics. For example, a violin made from the SMP will help to reduce the neck and shoulder pain

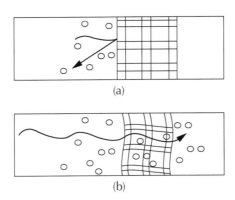

(a)

(b)

FIGURE 8.26 Schematic representation of molecular structure of fabric membrane at (a) below activation point, and (b) above activation point.

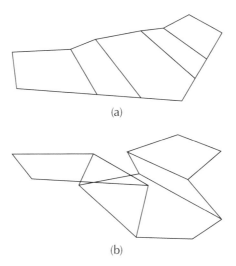

(a)

(b)

FIGURE 8.27   The wing can be adapted to different mission segments.

of the player because SMP can be changed to the player's demand. The potential market for the application of these materials is quite large. But, yet, only a few products have been commercialized in the world.

REFERENCES

www.aerofit.com.

www.diaplex.com.

Ahir, S. V., and E. M. Terentjev. 2005. Photomechanical actuation in polymer-nanotube composites. *Nature Materials* 4 (6):491-495.

Alteheld, A., Y. K. Feng, S. Kelch, and A. Lendlein. 2005. Biodegradable, amorphous copolyester-urethane networks having shape-memory properties. *Angewandte Chemie-International Edition* 44 (8): 1188–92.

Behl, M., and A. Lendlein. 2007. Shape-memory polymers. *Materials Today* 10 (4): 20–28.

Beloshenko, V. A., Y. E. Beygelzimer, A. P. Borzenko, and V. N. Varyukhin. 2002. Shape memory effect in the epoxy polymer-thermoexpanded graphite system. *Composites, Part A: Applied Science and Manufacturing* 33A (7): 1001–6.

Bhamra, T., and B. Hon. 2004. Design and manufacture for sustainable development, 2004. September 2004 at Burleigh Court, Loughborough University, UK. Bury St. Edmunds: Professional Engineering Pub.

Borden, T. 1991. Shape-memory alloys—Forming a tight fit. *Mechanical Engineering* 113 (10): 67–72.

Buckley, C. P., C. Prisacariu, and A. Caraculacu. 2007. Novel triol-crosslinked polyurethanes and their thermorheological characterization as shape-memory materials. *Polymer* 48 (5): 1388–96.

Buckley, P. R., G. H. McKinley, T. S. Wilson, W. Small, W. J. Benett, J. P. Bearinger, M. W. McElfresh, and D. J. Maitland. 2006. Inductively heated shape memory polymer for the magnetic actuation of medical devices. *IEEE Transactions on Biomedical Engineering* 53 (10): 2075–83.

Campbell, D. 2005. Elastic memory composite material: An enabling technology for future furable space structures. *46th AIAA/ASME/ASCE/AHS/ASC Structures, Structural Dynamics, and Materials Conference,* Austin, Texas.

Cao, F., and S. C. Jana. 2007. Nanoclay-tethered shape memory polyurethane nanocomposites. *Polymer* 48 (13): 3790–3800.

Chang, C. Y., D. Vokoun, and C. T. Hu. 2001. Two-way shape memory effect of NiTi alloy induced by constraint aging treatment at room temperature. *Metallurgical and Materials Transactions a-Physical Metallurgy and Materials Science* 32 (7): 1629–34.

Chang, R., and R. J. Nikolai. 1994. Temperature influences on nickel-titanium-alloy-wire responses in flexure. *Journal of Dental Research* 73:323.

Charlesby, A. 1960. Atomic radiation and polymers. New York: Pergamon Press, 198.

Chen, W., C. Y. Zhu, and X. R. Gu. 2002. Thermosetting polyurethanes with water-swollen and shape memory properties. *Journal of Applied Polymer Science* 84 (8): 1504–12. Reprinted with permission of John Wiley & Sons, Inc.

Chernous, D. A., S. V. Shil'ko, and Y. M. Pleskachevskii. 2004. Description of the shape effect of radiation polymers under thermomechanical action. *J. Eng. Phys. Thermophys.* 77:6.

Cho, J. W., Y. C. Jung, B. C. Chun, and Y. C. Chung. 2004. Water vapor permeability and mechanical properties of fabrics coated with shape-memory polyurethane. *Journal of Applied Polymer Science* 92 (5): 2812–16. Reprinted with permission of John Wiley & Sons, Inc.

Cho, J. W., Y. C. Jung, Y. C. Chung, and B. C. Chun. 2004. Improved mechanical properties of shape-memory polyurethane block copolymers through the control of the soft-segment arrangement. *Journal of Applied Polymer Science* 93 (5): 2410–15.

Chun, B. C., T. K. Cho, and Y. C. Chung. 2007. Blocking of soft segments with different chain lengths and its impact on the shape memory property of polyurethane copolymer. *Journal of Applied Polymer Science* 103 (3): 1435–41. Reprinted with permission of John Wiley & Sons, Inc.

Clowes, S. 2001. Smart Shirt rolls up its sleeves. http://new.bbc.co.uk.

Ding, X. M., J. L. Hu, and X. A. Tao. 2004. Effect of crystal melting on water vapor permeability of shape-memory polyurethane film. *Textile Research Journal* 74 (1): 39–43.

Fakirov, S. 2005. *Handbook of condensation thermoplastic elastomers.* Weinheim: Wiley-VCH.

Fuentes, J. M. G., P. Guempel, and J. Strittmatter. 2002. Phase change behavior of nitinol shape memory alloys: Influence of heat and thermomechanical treatments. *Advanced Engineering Materials* 4 (7): 437–51.

Gall, K., M. L. Dunn, Y. P. Liu, D. Finch, M. Lake, and N. A. Munshi. 2002. Shape memory polymer nanocomposites. *Acta Materialia* 50 (20): 5115–26.

Gall, K., M. L. Dunn, Y. P. Liu, G. Stefanic, and D. Balzar. 2004. Internal stress storage in shape memory polymer nanocomposites. *Applied Physics Letters* 85 (2): 290–92.

Gall, K., P. Kreiner, D. Turner, and M. Hulse. 2004. Shape-memory polymers for microelectromechanical systems. *Journal of Microelectromechanical Systems* 13 (3): 472–83.

Goethals, E. J., W. Reyntjens, and S. Lievens. 1998. Poly(vinyl ethers) as building blocks for new materials. *Macromolecular Symposia* 132: 57–64.

Harris, K. D., C. W. M. Bastiaansen, and D. J. Broer. 2007. Physical properties of anisotropically swelling hydrogen-bonded liquid crystal polymer actuators. *Journal of Microelectromechanical Systems* 16 (2): 480–88.

Havens, E., E. A. Snyder, and T. H. Tong. 2005. Light-activated shape memory polymers and associated applications. *Proceedings of SPIE—The International Society for Optical Engineering* 5762:48–55.

Hayashi, S. 1993. Properties and applications of polyurethane-series shape memory polymer. *Int. Progress in Urethanes* 6.

Hu, J., X. Ding, X. Tao, and J. Yu. 2002. Shape memory polymers and their applications to smart textile products. *Journal of Dong Hua University*, English Edition 19:89.

Hu, J. L., F. L. Ji, and Y. W. Wong. 2005. Dependency of the shape memory properties of a polyurethane upon thermomechanical cyclic conditions. *Polymer International* 54 (3): 600–605. Reprinted with permission of John Wiley & Sons, Inc.

Hu, J. L., Z. H. Yang, L. Y. Yeung, F. L. Ji, and Y. Q. Liu. 2005. Crosslinked polyurethanes with shape memory properties. *Polymer International* 54 (5): 854–59.

Ikematsu, T., Y. Kishimoto, and M. Karaushi. 1990. Block copolymer bumpers with good shape memory. Japan Patent. 02022355.

Inoue, K., S. I. Y. Shen, M. Taya, and National Science Foundation (U.S.). 1997. US-Japan Workshop on Smart Materials and Structures. *Proceedings of the First US-Japan Workshop on Smart Materials and Structures*, University of Washington, Seattle, Washington, December 1995. Warrendale, PA: Minerals, Metals & Materials Society.

Jeong, H. M., B. K. Ahn, S. M. Cho, and B. K. Kim. 2000. Water vapor permeability of shape memory polyurethane with amorphous reversible phase. *Journal of Polymer Science Part B-Polymer Physics* 38 (23): 3009–17.

Jeong, H. M., B. K. Ahn, and B. K. Kim. 2000. Temperature sensitive water vapour permeability and shape memory effect of polyurethane with crystalline reversible phase and hydrophilic segments. *Polymer International* 49 (12): 1714–21.

———. 2001. Miscibility and shape memory effect of thermoplastic polyurethane blends with phenoxy resin. *European Polymer Journal* 37 (11): 2245–52.

Jeong, H. M., S. Y. Lee, and B. K. Kim. 2000. Shape memory polyurethane containing amorphous reversible phase. *Journal of Materials Science* 35 (7): 1579–83.

Jeong, H. M., J. H. Song, S. Y. Lee, and B. K. Kim. 2001. Miscibility and shape memory property of poly(vinyl chloride)/thermoplastic polyurethane blends. *Journal of Materials Science* 36 (22): 5457–63.

Kagami, Y., J. P. Gong, and Y. Osada. 1996. Shape memory behaviors of crosslinked copolymers containing stearyl acrylate. *Macromolecular Rapid Communications* 17 (8): 539–43.

Kelly, A., and P. T. Mather. 2005. Crystallization of POSS in a PEG-based multiblock polyurethane. *Materials Research Society Symposium Proceedings* 847:93–98.

Khonakdar, H. A., S. H. Jafari, S. Rasouli, J. Morshedian, and H. Abedini. 2007. Investigation and modeling of temperature dependence recovery behavior of shape-memory crosslinked polyethylene. *Macromolecular Theory and Simulations* 16 (1): 43–52.

Kim, B. K., and S. Y. Lee. 1996. Polyurethanes having shape memory effects. *Polymer* 37:5781.

Kim, B. K., S. Y. Lee, J. S. Lee, S. H. Baek, Y. J. Choi, J. O. Lee, and M. Xu. 1998. Polyurethane ionomers having shape memory effects. *Polymer* 39 (13): 2803–8.

Kim, B. K., and S. H. Paik. 1999. UV-curable poly(ethylene glycol)-based polyurethane acrylate hydrogel. *Journal of Polymer Science Part a-Polymer Chemistry* 37 (15): 2703–9.

Kim, B. K., Y. J. Shin, S. M. Cho, and H. M. Jeong. 2000. Shape-memory behavior of segmented polyurethanes with an amorphous reversible phase: The effect of block length and content. *Journal of Polymer Science Part B-Polymer Physics* 38 (20): 2652–57.

Kim, E. A., Jee, K. K., Kim, Y. B., and You, S. J. 2008. Coil spring having two-way shape memory effect and the fabrication method thereof, and adiabatic product using the same. W/0/2008/088197.

Kim, H. D., T. J. Lee, J. H. Huh, and D. J. Lee. 1999. Preparation and properties of segmented thermoplastic polyurethane elastomers with two different soft segments. *Journal of Applied Polymer Science* 73 (3): 345–52.

Kobayashi, K., and S. Hayashi. 1992. Woven fabric made of shape memory polymer. US Patent, 5128197.

Kumar, S., and M. V. Pandya. 1997. Thermally recoverable crosslinked polyethylene. *Journal of Applied Polymer Science* 64 (5): 823–29.

Kusy, R. P., and J. Q. Whitley. 1994. Thermal characterization of shape-memory polymer blends for biomedical implantations. *Thermochimica Acta* 243 (2): 253–63.

Lee, B. S., B. C. Chun, Y. C. Chung, K. I. Sul, and J. W. Cho. 2001. Structure and thermo-mechanical properties of polyurethane block copolymers with shape memory effect. *Macromolecules* 34 (18): 6431–37.

Lee, S. H., J. W. Kim, and B. K. Kim. 2004. Properties of shape memory polyurethanes having crosslinks in soft and hard segments. *Smart Materials and Structures* 13 (6): 1345–50.

Lendlein, A., H. Y. Jiang, O. Junger, and R. Langer. 2005. Light-induced shape-memory polymers. *Nature* 434 (7035): 879–82.

Lendlein, A., and S. Kelch. 2002. Shape-memory polymers. *Angewandte Chemie-International Edition* 41 (12): 2034–57.

Lendlein, A., S. Kelch, and K. Kratz. 2006. Plastics with programmed memory. *Kunstoffe* 2:54.

Lendlein, A., and R. Langer. 2002. Biodegradable, elastic shape-memory polymers for potential biomedical applications. *Science* 296 (5573): 1673–76.

Lendlein, A., A. M. Schmidt, M. Schroeter, and R. Langer. 2005. Shape-memory polymer networks from oligo(epsilon-caprolactone)dimethacrylates. *Journal of Polymer Science Part a-Polymer Chemistry* 43 (7): 1369–81.

Li, F., L. Qi, J. Yang, M. Xu, X. Luo, and D. Ma. 2000. Polyurethane/conducting carbon black composites: Structure, electric conductivity, strain recovery behavior, and their relationships. *Journal of Applied Polymer Science* 75 (1): 68–77.

Li, F. K., Y. Chen, W. Zhu, X. Zhang, and M. Xu. 1998. Shape memory effect of polyethylene nylon 6 graft copolymers. *Polymer* 39 (26): 6929–34.

Li, F. K., J. Hasjim, and R. C. Larock. 2003. Synthesis, structure, and thermophysical and mechanical properties of new polymers prepared by the cationic copolymerization of corn oil, styrene, and divinylbenzene. *Journal of Applied Polymer Science* 90 (7): 1830–38.

Li, F. K., J. N. Hou, W. Zhu, X. Zhang, M. Xu, X. L. Luo, D. Z. Ma, and B. K. Kim. 1996. Crystallinity and morphology of segmented polyurethanes with different soft-segment length. *Journal of Applied Polymer Science* 62 (4): 631–38. Reprinted with permission of John Wiley & Sons, Inc.

Li, F. K., and R. C. Larock. 2002. New soybean oil-styrene-divinylbenzene thermosetting copolymers. v. shape memory effect. *Journal of Applied Polymer Science* 84 (8): 1533–43.

Li, F. K., A. Perrenoud, and R. C. Larock. 2001. Thermophysical and mechanical properties of novel polymers prepared by the cationic copolymerization of fish oils, styrene and divinylbenzene. *Polymer* 42 (26): 10133–45.

Li, F. K., X. Zhang, J. N. Hou, M. Xu, X. L. Lu, D. Z. Ma, and B. K. Kim. 1997. Studies on thermally stimulated shape memory effect of segmented polyurethanes. *Journal of Applied Polymer Science* 64 (8): 1511–16. Reprinted with permission of John Wiley & Sons, Inc.

Li, F. K., W. Zhu, X. Zhang, C. T. Zhao, and M. Xu. 1999. Shape memory effect of ethylene-vinyl acetate copolymers. *Journal of Applied Polymer Science* 71 (7): 1063–70.

Liang, C., C. A. Rogers, and E. Malafeew. 1997. Investigation of shape memory polymers and their hybrid composites (Reprinted from Proceedings of the Second Joint Japan/US Conference on Adaptive Structures, Nov, pp. 789–802). *Journal of Intelligent Material Systems and Structures* 8 (4): 380–86.

Lim, L. T., I. J. Britt, and M. A. Tung. 1999. Sorption and transport of water vapor in nylon 6,6 film. *Journal of Applied Polymer Science* 71 (2): 197–206.

Lin, J. R., and L. W. Chen. 1998a. Study on shape-memory behavior of polyether-based polyurethanes. I. Influence of the hard-segment content. *Journal of Applied Polymer Science* 69 (8): 1563–74.

————. 1998b. Study on shape-memory behavior of polyether-based polyurethanes. II. Influence of soft-segment molecular weight. *Journal of Applied Polymer Science* 69 (8): 1575–86.

————. 1999. Shape-memorized crosslinked ester-type polyurethane and its mechanical viscoelastic model. *Journal of Applied Polymer Science* 73 (7): 1305–19. Reprinted with permission of John Wiley & Sons, Inc.

Liu, C., S. B. Chun, P. T. Mather, L. Zheng, E. H. Haley, and E. B. Coughlin. 2002. Chemically cross-linked polycyclooctene: Synthesis, characterization, and shape memory behavior. *Macromolecules* 35 (27): 9868–74.

Liu, C., and P. T. Mather. 2002. Thermomechanical characterization of a tailored series of shape memory polymers. *Journal of Applied Medical Polymers* 6 (2): 47–52.

Liu, C., H. Qin, and P. T. Mather. 2007. Review of progress in shape-memory polymers. *Journal of Materials Chemistry* 17 (16): 1543–58.

Liu, G. Q., C. L. Guan, H. S. Xia, F. Q. Guo, X. B. Ding, and Y. X. Peng. 2006. Novel shape-memory polymer based on hydrogen bonding. *Macromolecular Rapid Communications* 27 (14): 1100–1104.

Liu, Y., K. Gall, M. L. Dunn, A. R. Greenberg, and J. Diani. 2006. Thermomechanics of shape memory polymers: Uniaxial experiments and constitutive modeling. *International Journal of Plasticity* 22 (2): 279–313.

Liu, Y., K. Gall, M. L. Dunn, and P. McCluskey. 2004. Thermomechanics of shape memory polymer nanocomposites. *Mechanics of Materials* 36:929.

Lu, T. J., and A. G. Evans. 2002. Design of a high authority flexural actuator using an electrostrictive polymer. *Sensors and Actuators, A: Physical* A99 (3): 290–96.

Luo, X. L., X. Y. Zhang, M. T. Wang, D. H. Ma, M. Xu, and F. K. Li. 1997. Thermally stimulated shape-memory behavior of ethylene oxide ethylene terephthalate segmented copolymer. *Journal of Applied Polymer Science* 64 (12): 2433–40.

Mather, P. T., H. G. Jeon, and T. S. Haddad. 2000. Strain recovery in POSS hybrid thermoplastics. *Polymer Preprints (American Chemical Society, Division of Polymer Chemistry)* 41 (1): 528–29.

Mather, P. T., B. S. Kim, Q. Ge, and C. Liu. 2004. Synthesis of nonionic telechelic polymers incorporating polyhedral oligosilsesquioxane and uses thereof. US Patent, 2004024098.

Mather, P. T., and C. Liu. 2003. Castable shape memory polymers. World Patent, WO 2003093341.

Mather, P. T., H. Qin, J. Wu, and J. Bobiak. 2006. POSS-based polyurethanes: From degradable polymers to hydrogels. *Medical Polymers 2006 5th,* Cologne, Germany, RAPRA, Shrewsbury, UK.

Mattila, H. R. 2006a. *Intelligent textiles and clothing.* Chapter 6. Boca Raton, FL: CRC Press.

————. 2006b. *Intelligent textiles and clothing.* Chapter 7. Boca Raton, FL: CRC Press.

Metcalfe, A., A. C. Desfaits, I. Salazkin, L. Yahia, W. M. Sokolowski, and J. Raymond. 2003. Cold hibernated elastic memory foams for endovascular interventions (vol. 24, pg. 491, 2003). *Biomaterials* 24 (9): 1681.

Mirfakhrai, T., J. D. W. Madden, and R. H. Baughman. 2007. Polymer artificial muscles. *Materials Today* 10 (4): 30–38.

Mondal, S., and J. L. Hu. 2006. Temperature stimulating shape memory polyurethane for smart clothing. *Indian Journal of Fibre and Textile Research* 31 (1): 66–71.

————. 2007. Studies of shape memory property on thermoplastic segmented polyurethanes: Influence of PEG 3400. *J. Elast. Plast.* 39:81.

Mullins, W. S., M. D. Bagby, and T. L. Norman. 1996. Mechanical behavior of thermo-responsive orthodontic archwires. *Dental Materials* 12:308–14.

Naga, N., G. Tsuchiya, and A. Toyota. 2006. Synthesis and properties of polyethylene and polypropylene containing hydroxylated cyclic units in the main chain. *Polymer* 47 (2): 520–26.

Nagai, H., A. Ueda, and S. Isomura. 1994. Shape-memory Norbornene Polymer Molded Products. Japan Patent. 06080768.

Nagata, M., and I. Kitazima. 2006. Photocurable biodegradable poly(.vepsiln.-caprolactone)/ poly(ethylene glycol) multiblock copolymers showing shape-memory properties. *Colloid and Polymer Science* 284 (4): 380–86.

Newnham, R. E. 1998. Phase transformations in smart materials. *Acta Crystallographica, Section A: Foundations of Crystallography* A54 (6): 729–37.

Ni, X. Y., and X. H. Sun. 2006. Block copolymer of trans-polyisoprene and urethane segment: Shape memory effects. *Journal of Applied Polymer Science* 100 (2): 879–85.

Otsuka, K., and C. M. Wayman. 1998. *Mechanism of shape memory effect and superelasticity, shape memory materials.* Cambridge: Cambridge University Press, 27–48.

Park, H. S., J. W. Kim, S. H. Lee, and B. K. Kim. 2004. Temperature-sensitive amorphous polyurethanes. *Journal of Macromolecular Science, Part B: Physics* 43 (2): 447–58.

Ping, P., W. S. Wang, X. S. Chen, and X. B. Jing. 2005. Poly(epsilon-caprolactone) polyure-thane and its shape-memory property. *Biomacromolecules* 6 (2): 587–92.

Qin, H., and P. T. Mather. 2006. Polyurethane thermoplastics containing polyhedral oligo-meric silsesquioxane (POSS) units. *PMSE Preprints* 94:127–28.

Reyntjens, W. G., F. E. Du Prez, and E. J. Goethals. 1999. Polymer networks contain-ing crystallizable poly(octadecyl vinyl ether) segments for shape-memory materials. *Macromolecular Rapid Communications* 20 (5): 251–55.

Rousseau, I. A. 2004. Development of soft polymeric networks showing actuation behavior: From hydrogels to liquid crystalline elastomers. Thesis, University of Connecticut.

———. 2008. Challenges of shape memory polymers: A review of the progress toward over-coming SMP's limitations. *Polymer Engineering and Science* 48 (11): 2075–89.

Rousseau, I. A., and P. T. Mather. 2003. Shape memory effect exhibited by smectic-C liq-uid crystalline elastomers. *Journal of the American Chemical Society* 125 (50): 15300–15301.

Sahoo, N. G., Y. C. Jung, N. S. Goo, and J. W. Cho. 2005. Conducting shape memory polyure-thane-polypyrrole composites for an electroactive actuator. *Macromolecular Materials and Engineering* 290 (11): 1049–55.

Shirai, Y., and S. Hayashi. 1988. Development of polymeric shape memory material. MTB184, Mitsubishi Heavy Industries, Inc., Japan, December.

Skakalova, V., V. Lukes, and M. Breza. 1997. Shape memory effect of dehydrochlorinated crosslinked poly(vinyl chloride). *Macromolecular Chemistry and Physics* 198 (10): 3161–72.

Sokolowski, W. 2005. Potential biomedical and commercial applications of cold hibernated elastic memory (CHEM) self-deployable foam structures. *Proceedings of SPIE-The International Society for Optical Engineering.* 5648:397–405.

Sokolowski, W., Metcalfe, A., and Hayashi. 2007. Medical application of dhape memory poly-mers. *Biomedical Materials* 2 (1).

Sokolowski, W. M., and A. B. Chmielewski. 1999. Cold hibernated elastic memory (CHEM) self-deployable structures. *Proceedings of SPIE-The International Society for Optical Engineering* 3669:179–85.

Sun, X. H., and X. Y. Ni. 2004. Block copolymer of trans-polyisoprene and urethane segment: Crystallization behavior and morphology. *Journal of Applied Polymer Science* 94 (6): 2286–94.

Takahashi, T., N. Hayashi, and S. Hayashi. 1996. Structure and properties of shape-memory polyurethane block copolymers. *Journal of Applied Polymer Science* 60 (7): 1061–69.

Tang, W., B. Sundman, R. Sandstrom, and C. Qiu. 1999. New modelling of the B2 phase and its associated martensitic transformation in the Ti-Ni system. *Acta Materialia* 47 (12): 3457–68.

Tobushi, H., S. Hayashi, A. Ikai, and H. Hara. 1996. Thermomechanical properties of shape memory polymers of polyurethane series and their applications. *Journal De Physique Iv* 6 (C1): 377–84.

Tong, T. H. 2002. Shape memory styrene copolymer. World Patent, WO 2002059170.

Vaia, R. 2005. Nanocomposites—Remote-controlled actuators. *Nature Materials* 4 (6): 429–30.

Vili, Y. Y. F. C. 2007. Investigating smart textiles based on shape memory materials. *Textile Research Journal* 77 (5): 290–300.

Wache, H. M., D. J. Tartakowska, A. Hentrich, and M. H. Wagner. 2003. Development of a polymer stent with shape memory effect as a drug delivery system. *Journal of Materials Science-Materials in Medicine* 14 (2): 109–12.

Wang, H. H., and U. E. Yuen. 2006. Synthesis of thermoplastic polyurethane and its physical and shape memory properties. *Journal of Applied Polymer Science* 102 (1): 607–15.

Wang, M. T., X. L. Luo, X. Y. Zang, D. Z. Ma, and L. D. Zang. 1997. Shape memory properties in poly(ethylene oxide)-poly(ethylene terephthalate). *Polymers for Advanced Technologies* 8 (136).

Wang, M. T., and L. D. Zhang. 1999. Recovery as a measure of oriented crystalline structure in poly(ether ester)s based on poly(ethylene oxide) and poly(ethylene terephthalate) used as shape memory polymers. *Journal of Polymer Science Part B-Polymer Physics* 37 (2): 101–12.

Wang, W. S., P. Ping, X. S. Chen, and X. B. Jing. 2006. Polylactide-based polyurethane and its shape-memory behavior. *European Polymer Journal* 42 (6): 1240–49.

———. 2007. Biodegradable polyurethane based on random copolymer of L-lactide and epsilon-caprolactone and its shape-memory property. *Journal of Applied Polymer Science* 104 (6): 4182–87.

Wang, Y. L., Y. Hu, X. L. Gong, W. Q. Jiang, P. Q. Zhang, and Z. Y. Chen. 2007. Preparation and properties of magnetorheological elastomers based on silicon rubber/polystyrene blend matrix. *Journal of Applied Polymer Science* 103 (5): 3143–49.

Wei, Z. G., R. Sandstrom, and S. Miyazaki. 1998. Shape-memory materials and hybrid composites for smart systems. Part I. Shape-memory materials. *Journal of Materials Science* 33 (15): 3743–62.

Wilson, T. S., W. I. V. Small, W. J. Benett, J. P. Bearinger, and D. J. Maitland. 2005. Shape memory polymer therapeutic devices for stroke. *Proceedings of SPIE-The International Society for Optical Engineering*. 6007:60070R/1 - 60070R/8.

Xu, J. W., W. F. Shi, and W. M. Pang. 2006. Synthesis and shape memory effects of Si-O-Si cross-linked hybrid polyurethanes. *Polymer* 47 (1): 457–65.

Yang, B., W. M. Huang, C. Li, and L. Li. 2006. Effects of moisture on the thermomechanical properties of a polyurethane shape memory polymer. *Polymer* 47 (4): 1348–56.

Yang, F. Q., S. L. Zhang, and J. C. M. Li. 1997. Impression recovery of amorphous polymers. *Journal of Electronic Materials* 26 (7): 859–62.

Yang, J. H., B. C. Chun, Y. C. Chung, and J. H. Cho. 2003. Comparison of thermal/mechanical properties and shape memory effect of polyurethane block-copolymers with planar or bent shape of hard segment. *Polymer* 44 (11): 3251–58.

Yang, Z. H., J. L. Hu, Y. Q. Liu, and L. Y. Yeung. 2006. The study of crosslinked shape memory polyurethanes. *Materials Chemistry and Physics* 98 (2-3): 368–72.

Zhu, G., G. Liang, Q. Xu, and Q. Yu. 2003. Shape-memory effects of radiation crosslinked poly(epsilon-caprolactone). *Journal of Applied Polymer Science* 90 (6): 1589–95.

Zhu, G. M., Q. Y. Xu, G. Z. Liang, and H. F. Zhou. 2005. Shape-memory behaviors of sensitizing radiation-crosslinked polycaprolactone with polyfunctional poly(ester acrylate). *Journal of Applied Polymer Science* 95 (3): 634–39.

Zhu, G. M., S. G. Xu, J. H. Wang, and L. B. Zhang. 2006. Shape memory behaviour of radiation-crosslinked PCL/PMVS blends. *Radiation Physics and Chemistry* 75 (3): 443–48.

9 Methods of Evaluation for Wearable Computing

Daniel Ashbrook, Kent Lyons,
James Clawson, and Thad Starner

CONTENTS

9.1   INTRODUCTION

With the rise of smart textiles, wearable computing is poised to enter mainstream culture. Wearables are not only used in isolation, but alongside of and in cooperation with other technologies, both by individuals and in social settings, both in quiet, stationary situations and while "on-the-go." Many of the attributes that make wearable computing compelling also make for a difficult environment in which to conduct rigorous research and evaluations. Unlike traditional desktop computers, wearable technologies encourage the user to interact with the technology any place, any time, and for any duration. The anywhere, anytime nature of interaction with wearables makes it possible to form intimate bonds with the technology, such that it becomes meaningful to consider a "single use" as taking place over a range of time, spanning from a sub-second burst to an interaction that takes place over many years.

In order to effectively conduct research in this domain, it is important to understand the potential impact of continuous use in everyday environments and how this usage may affect both the wearable technologies and their users. However, it can be quite difficult to conduct standard usability studies in this domain; therefore, one major challenge is that of methodology: what are appropriate methods for conducting evaluations of wearable technologies, and how should such studies be designed? For nearly 15 years, we have performed research in this area, drawing both on a significant body of knowledge gained from using wearable technologies in everyday life (combined, the authors have over 20 man-years of such firsthand experience), and on an ever-expanding assortment of formal evaluations designed to uncover specific phenomena related to the wearable experience. While our past work has focused on creating devices and sensing systems for new mobile interfaces, our current work has expanded to involve the creation of novel textile interfaces. We are drawing upon our past experiences to evaluate our new multitouch textile interface (described below), attempting to demonstrate its use and usefulness. While some of our user studies are ongoing, we include them to illustrate how the evaluation methods presented here can be adapted for textile interfaces.

In this chapter, we discuss several approaches for formally evaluating wearable computing. We have used these techniques to understand usability issues related to body-worn technologies being utilized by able-bodied users. Together, these methods provide a framework for researchers and scientists interested in conducting user-centered evaluations of technologies worn on the body or embedded into smart textiles.

9.1.1 INTERNAL AND ECOLOGICAL VALIDITY

The research methods we present here fall along a methodological continuum (Figure 9.1). On one extreme are rigorously controlled laboratory-based studies, involving careful study design focused on examining one or two specific factors of interest. In order to isolate the desired variables, these studies must often be performed in a fixed (non-mobile) laboratory setting that bears little resemblance to the everyday situations where the technology under evaluation will actually be used. Often such studies seek to be internally valid in that they may allow one to posit a causal connection between a changed, independent parameter and an outcome.

At the opposite extreme are studies conducted "in the wild." These evaluations are intended to examine behavior and technology use in the everyday setting of the participants, rather than to extract reproducible statistical knowledge. Highly

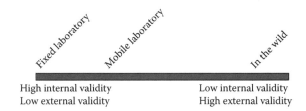

FIGURE 9.1   The methodological continuum of validity.

ecologically valid,* such studies offer the most realistic setting for understanding a given technology; however, they also provide many challenges. The real-life nature of everyday settings often causes difficulties in collecting data, and analyzing the collected data can be even more problematic with respect to internal validity. Because of the lack of experimental controls, these studies require careful planning on how usage information will be captured, and the researcher may have difficulty in uncovering and interpreting results from the data.

Between these two extremes on the methodological continuum of validity are hybrid techniques that try to strike a balance between using experimental controls to answer quantitative questions reproducibly, and gaining an understanding of how everyday phenomena might impact users and their technology. Below, we present the methods used for several studies that fall along this continuum. We discuss relevant aspects of the design of the experiments and the benefits and limitations of each. Taken together, they provide a framework for conducting new evaluations and designing new methods for evaluating wearable technologies. Finally, we discuss some practical considerations that should be taken into account when conducting such studies of wearable technologies, including data collection and privacy concerns.

9.1.2 GENERAL STUDY DESIGN AND TOOLS

Typically, quantitative studies are performed either between subjects or within subjects. Within-subject designs often compare two or more types of interfaces used by each participant. Each participant is therefore tested in each experimental condition. The order of the conditions is randomized or counterbalanced to minimize the impact of potential learning effects (i.e., where the use of the first device may teach the subject how to better use subsequent ones) or other types of transfer effects.

Between-subject studies often reserve one interface, design, or more generally, experimental condition for each group of participants. There could be several reasons to use such a study design: the researcher suspects that learning effects will overwhelm the comparative effect of the interface, the study requires too much time or effort per condition for each participant to complete all conditions, or there are too many conditions. For example, with consumer devices, many companies are concerned with novice use as they believe that a potential buyer must be able to use a demonstration model within a few minutes. In this case, if one is trying to choose between two novel, but similar, interfaces for a new product to optimize a novice's experience, the researcher cannot conduct a within-subjects study as the user's experience with the first device will contaminate their perception of the second device. For more information on experiment design, please refer to a modern human-computer interaction textbook (Dix et al. 2004).

It is important, before conducting an experiment, to decide what is to be measured. Table 9.1 shows a list of metrics we typically use in our experiments on mobile and wearable interfaces. All of the metrics listed for fixed in-the-lab studies may be used for mobile in-the-lab studies. Similarly, for in-the-wild studies, the metrics

---

* Note, ecological validity is different from external validity, but is often related. External validity refers to how well the results from a study generalize to other situations.

TABLE 9.1

A Selection of Metrics for Usability Experiments for Mobile Interfaces

Fixed in-the-Lab

Time required to use the interface

Accuracy

Access time

Learning curve

NASA Task Load Index (TLX)

Likert scale questions

Eye tracking (time looking at interface vs. environment)

Hand usage (which hand used, motion tracking, etc.)

Mobile in-the-Lab

Footstep rate

Distance traveled

Velocity

Travel accuracy

In-the-Wild

Usage

Time on task

Balking

from both fixed and mobile in-the-lab studies might be used. However, as the user becomes more mobile, sensing and storing the data for these metrics becomes difficult, and the user may have to wear more bulky and difficult to maintain equipment. The requirements for internal and ecological validity are often at odds with mobile experiments. The more extraneous equipment the subject has to wear for the experiment, the more it will affect the use of the device.

While the metrics chosen will depend on the parameters of the specific experiment, the types of metrics listed in the table usually relate strongly to the user's experience. For example, the time required to use an interface and the participants' accuracy are often key metrics for their perceived performance. If an interface is designed for use while the user is performing another task (e.g., changing the track on an MP3 player while reading e-mail), the relative effect of the interface on both tasks can often be measured in terms of time and accuracy. Going further, the user's ability to learn an interface (the learning curve) is often expressed in the amount of time the user is exposed to the interface versus the speed and accuracy of use. Thus, learning curves show the change in user performance over time. In our experience, the amount of time needed to acquire a mobile device (e.g., retrieving a mobile phone from a pocket or bag) and begin using the interface is often critical to the success of that interface (Ashbrook et al. 2008). For convenience, we call this quantity "access time."

Combining access time, learning curves, and dual task situations may be of particular interest for designers of textile interfaces. An advantage of many textile interfaces is that they can be integrated with the user's clothing, allowing fast access to the interface and use while the user is focused on another (dual) task. One can imagine a researcher creating a textile interface for the control of a mobile phone while walking (e.g., a mobile phone keypad embroidered on an armband or backpack strap). The researcher might design an evaluation that compares the use of the textile interface to a normal handset while the subject walks a pre-defined path. Measuring average access time should immediately show one advantage of the textile interface (i.e., the time required to retrieve the handset interface from the pocket is not needed for the textile interface). However, due to the subjects' inexperience with a textile interface, the users might be less accurate and take more time for a given interaction on the textile interface at first. A comparison of the learning curves between the two conditions should show that the subject quickly becomes more proficient at using the textile interface.

While the efficiency of an interface is important, users' subjective experiences can often drive adoption of an interface. The NASA Task Load Index (TLX) (Hart and Staveland 1988) is a means of quantifying subjective effects of a user's overall workload and is well established in the literature. Furthermore, it has several subscales including mental demand, physical demand, temporal demand (time pressure), effort (how hard the task was), performance (how well the participant felt they did), and frustration level, which can often illuminate unexpected benefits or problems with an interface.

Surveys are another common method of eliciting subjective user experiences. Likert scale questions are often used in survey design. Here, the researcher posits a statement such as, "I could determine the location of the embroidered buttons easily," and provides a scale, 1 to 7, with 1 being "strongly disagree," 4 "neither agree nor disagree," and 7 "strongly agree." The user circles the number that most matches his or her sentiment (in general, the literature shows that a seven-number scale works well). We have also found that providing a free response area for participants at the end of a qualitative survey is wise. Simply having a final question of, "Is there anything else you would like to share with the researchers?" can lead to unexpected insights and topics for further study. These questions also enable a transition into a more open-ended, semi-structured interview with the participants, when appropriate.

Mobile metrics often focus on the implicit dual task imposed by mobility, such as walking or driving. Even using a mobile device while riding a bus or subway is a dual task situation in that the user must constantly monitor the progression of stops and others' movement around him. Speed and accuracy of movement are often compared between a condition where the mobile interface is used and a condition where it is not used. Thus, the effect of the use of the device on the person's ability to move is established. Conversely, the use of the interface while stationary can be compared to its use while moving so that the effect of motion on the interface can be determined. Dual task metrics, when combined with more subjective measures such as the NASA TLX, provide the researcher with a sense of the amount of interference occurring between the two tasks.

In-the-wild experiments can, in theory, use all of the metrics discussed above. However, the burden of mobile instrumentation is often too high in practice. More often researchers deploy more mature interfaces with a group of users appropriate for their target population and record their usage patterns. The most straightforward metric is how much the interface is used, and often the interface software can be modified to record this information. If the user's environment is instrumented, typically with video cameras, more information can be gathered. For example, if the interface involves scheduling appointments during meetings, the researchers might review the video recorded in an office environment at the time the interface was used to determine how much time was spent on the interface as opposed to the verbal negotiation of the appointment time. This "time on task" metric may reveal how to better streamline interfaces for real-world situations.

Another in-the-wild metric we are attempting to define is balking. The term is from queueing theory where it refers to situations where customers attempt to use a service, find it busy, and leave without completing their business. Here, we use the term to refer to situations where using a mobile interface would make sense logically, but the participant does not use it. We have found balking to be a particular problem with mobile calendaring applications (Starner et al. 2004). Access time and a lack of synchronization with the user's main scheduling device (often a desktop PC) seemed major difficulties for the interfaces. In our observation, face-to-face conversation incurs a social pressure to keep the flow of conversation moving, which inhibits the use of all but the fastest to access mobile interfaces (e.g., glancing at a wristwatch to determine the time). Creating socially acceptable mobile interfaces is a challenge, and a measure of balking would seem to be a useful metric. However, recording the user's in-the-wild experience and determining every situation where he or she might have used the interface (and did not) is onerous. Another approach could be to reward the user for using the interface upon hearing an audio or vibration cue from his or her mobile device, triggered at random times throughout the day (Hudson et al. 2002). In this manner, the researcher can better index recordings of the subject's life to examine his or her behavior. Another tactic is to use a similar approach to context-aware experience sampling (Intille et al. 2003) and have the alert occur when a set of contextual conditions are met (for example, only alert the user to use the interface when he or she is in conversation to explore how interruptive different interfaces are). Such techniques can greatly save time on reviewing the data from an in-the-wild participant.

9.2   FIXED IN-THE-LABORATORY STUDIES

In order to further their goals, researchers often require reproducible, high-quality, quantitative data about a practice, technology, or interaction technique. There are many reasons why a non-mobile, in-the-lab study may be the best technique for gathering such results. Frequently, researchers need to understand existing practices before trying to improve on them or want to get an idea of baseline performance for some task before designing further research around it. Many times, studies are run to choose between several alternative designs or to justify that further, more comprehensive studies are warranted to explain a particular effect. Sometimes the data collection task is too complex or difficult to perform in a real-life situation or laboratory data is sufficient without the need to explore the full spectrum of everyday use.

Often, laboratory studies take place while the user is stationary, seated, and in a quiet environment away from external distractions. Considerable effort is taken to eliminate random sources of influence that might affect the results of the study, such as noise, other people, or changing visual stimuli. The laboratory nature of these studies affords the evaluation of early prototypes and mock-ups as well as more polished versions of wearable systems. The laboratory setting also permits the device under study to be connected to a desktop computer to facilitate data collection or running the experimental procedure. In some circumstances, it may be desirable to record the actions of the participant, using screen recording software or video cameras. Often such recordings are useful to investigate unusual results or to determine why some subjects performed abnormally well, or poorly, compared to the average. In other situations, the recorded media may be coded by the researchers to create a quantitative record of key events during the study.

9.2.1 TOUCHSCREEN WATCH STUDY

We have conducted a number of fixed in-the-laboratory studies of interfaces designed for mobility. The first example we will discuss is a project exploring interface options for round touchscreen wristwatches (Ashbrook, Lyons, and Starner 2008). Because touch interactions with small, round displays had not been widely reported in the literature, we wanted to understand the best-case performance that could be expected of users before prototyping novel interactions or exploring possible effects from in-situ usage. To this end, we conducted an experiment to investigate the size of virtual buttons needed to enable finger-based interaction on a small, round touchscreen watch. In particular, we wanted to understand the trade-offs between the number of software buttons placed around the edge of the display, the amount of non-button area remaining in the center, and the error rate for each of three types of movement. The results from this evaluation represent the upper limit of performance and inform designers how to start thinking about the best way to create interactions around this limit.

Because the hardware that we envision for our interface designs does not currently exist, we created a tethered mock-up. The participant wore a modified smartphone on his or her wrist, which was connected to a host computer that controlled the study parameters and logged all data. Although tethering was not strictly necessary, it greatly simplified the mechanics of the study and maintained high reliability in terms of data collection. Tethering the hardware was an option because the study was non-mobile. We felt that, due to the very preliminary nature of the investigation, running the study with mobile conditions would be premature and might compromise the validity of the results. Therefore, the study was designed to focus on internal validity rather than ecological validity.

9.2.2 Text Entry Study

For other experiments, additional ecological validity might be gained by simply modifying the parameters of the experiment. One example is our research on text entry for wearable computers. From our personal experience, we knew that some wearable computer users can use the one-handed Twiddler chording keyboard (Figure 9.2) to rapidly enter text in a wide variety of settings. In initial formal experiments, we systematically trained novices to type on the keyboard while seated. With sufficient practice, the subjects were able to attain a high rate of text entry (Lyons, Plaisted, and Starner 2004). Similar studies showed that other text entry devices, such as mini-QWERTY keyboards (Figure 9.3), can offer equivalently fast expert typing rates (Clarkson et al. 2005).

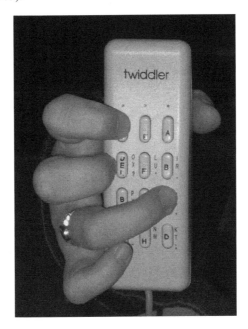

FIGURE 9.2   The Twiddler one-handed chording keyboard.

FIGURE 9.3   A typical mini-QWERTY keyboard.

Although the results from these studies were encouraging, they lacked ecological validity because they were performed in ideal conditions in the laboratory. Real wearable computer use takes place in a wide variety of settings, including those where the user cannot focus all of his or her attention on the keyboard during text entry. To get an idea of how users' performance might change during such conditions, we conducted another laboratory experiment using our expert typists from the prior studies. To simulate conditions of limited visual attention, we experimentally manipulated visual feedback, removing participants' ability to see the keyboard, the output of their typing, or both.

These experiments show how effective even a slight modification of experimental procedure can be in increasing ecological validity. While the baseline studies showed roughly equivalent performance between the Twiddler and mini-QWERTY keyboards for experts, the modified study revealed significant differences. The Twiddler was just as effective with or without visual feedback. Typists denied visual feedback in the mini-QWERTY condition, on the other hand, suffered from a considerable increase in error rate that did not recover even with additional practice (Clarkson et al. 2005). By adding a small amount of ecological validity to this study along a particularly important dimension (visual attention), we revealed a significant difference for the two types of keyboards. The Twiddler remained effective, while the performance of the mini-QWERTY keyboard did not persist in less-than-ideal circumstances.

9.3   MOBILE IN-THE-LABORATORY STUDIES

As discussed in the last section, carefully designed stationary laboratory studies can often increase the ecological validity of results. However, in many cases this is not sufficient or not possible. In these circumstances, the researcher will likely sacrifice some internal validity to get the desired results. The question then is one of how much validity must be sacrificed, and the answer depends on the specific parameters of the study. In general, there tends to be a trade-off with higher ecological validity leading to lower internal validity; as the environment for a study more and more resembles real life, uncontrollable variables such as level of distraction, clothing type, or factors such as crowds or traffic become an issue. These variables are not detrimental to the study; on the contrary, they are the very factors that lend ecological validity to the experiment.

Therefore, a logical next step along the methodological continuum of validity is the mobile in-laboratory study. This variety of study reveals how wearable technology might be used while the user is in motion, but remains best-case with respect to other ecological factors such as distraction. Such studies have the participant walk naturally through a controlled environment while using the technology in question. Some previous research has utilized a treadmill to allow the user to walk (Barnard et al. 2005). In our research, we have tried to achieve further ecological validity by using a looped track around which the user walks while performing some task.

The benefits of a treadmill include little needed space and repeatability of experiments (many treadmills allow the user to input a program of speed and inclinations). However, by walking on a treadmill, study participants lose the ability to naturally control their walking pace and may focus attention on the treadmill itself, which could cause distractions not present in normal walking situations. Additionally, walking on a treadmill presents none of the obstacle avoidance or navigation tasks normally encountered while walking. Using a track immediately increases ecological validity as compared to a treadmill; the user is able to naturally control his or her walking pace and must engage in normal obstacle avoidance and navigation tasks. Researchers may also get an idea of how distracting the task under consideration is by comparing the participant's normal walking pace with the pace measured during the various experimental conditions. The downside of a track is that it occupies a larger amount of space, at least during periods of active experimentation, and depending on the design, may not yield results as repeatable as a treadmill.

There are three basic track designs that we have considered for mobile laboratory studies. The first is an on-floor track. Similar to an outdoor running track, this design uses taped or painted lines on the floor to delineate the sides of the track (Figure 9.4). Because the track edges are very clear, the researchers can keep a count of how many times the participant steps off of the track and use that measure as an indication of walking performance (Vadas et. al 2006). However, because the track is laid out on the floor, the participant must frequently look down to ensure that he or she keeps on the track. This action may decrease the ecological validity somewhat, as during natural walking conditions people tend to keep their heads raised to anticipate upcoming obstacles.

The next track design improves ecological validity by transforming walking from a head-down task to a head-up one. By placing flags or other markers at head level (Figure 9.5), participants can walk normally, using the flags as a guide. Ecological validity aside, the choice between flags and a taped or painted track is a practical consideration. A flag course can be harder to deploy, as it relies on a ceiling structure amenable to attachment, and can be harder to recreate for later studies, as hanging obstacles can present a hazard to re-using the space in between studies. Our solution to the latter problem is to use magnets glued to the flags and to the ceiling structure, so that the flags can be easily removed and replaced at will, while keeping their positions the same. Flag tracks also must be carefully designed to point participants in the right direction, especially in the case of tracks that overlap themselves (Figure 9.6).

The final design option for a track is to use a hallway or other naturally constrained environment. This design can be extremely easy to set up, often being only a matter of running studies after hours when other people have gone home, or of pushing around some furniture to create a pathway. The downside is that it can be hard to re-create the same path every time, or to ensure that every participant experiences the same track. If someone happens to walk across the participant's path in the hallway, it can invalidate that session of the experiment. This kind of track can confuse participants as to where to go if not well designed.

Data collection for mobile in-the-lab studies can be done either on the mobile participant's body, or with lab-based collection tools. Using wireless transmission, data can be collected on the participant and saved to a fixed computer in the lab. The laboratory also affords external recording of the mobile user. For instance, cameras can be placed strategically throughout the lab environment to record the participant as he or she moves along the path. While obviously not as simple as recording a user

FIGURE 9.4 The taped floor track used in the reading on-the-go study.

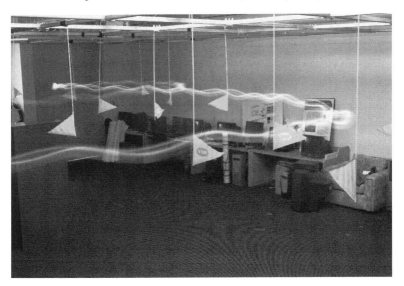

FIGURE 9.5 A flag track. A slow exposure was used with a flashlight to show the walking path.

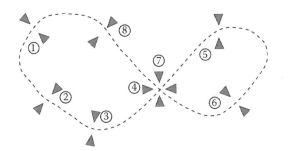

FIGURE 9.6   An overhead rendering of the track illustrated in Figure 9.5.

in a stationary lab study, the constrained environment makes such recordings possible. Researchers may also want to record information about the user's mobility. For instance, accelerometers can be placed on the user's body to record information about movement. The accelerometers might sense the user's step rate, head orientation relative to ground, or limb movement. Like the other recorded data, this information can be recorded on-body, or sent to a laboratory computer over a wireless link.

While useful, these devices cannot recover the subject's absolute position on the track. Additional sensors must be used. Recently, advances in inertial navigation systems (INS) have made it possible to track a user with relatively good precision using a combination of body-worn sensors without external infrastructure. See Beauregard's PhD dissertation "Infrastructureless Pedestrian Positioning" (Beauregard 2009) for a review of current commercial and research systems. If the experimental track is outdoors, on-body global positioning system (GPS) receivers may also be used to increase the accuracy of tracking, but consumer devices often do not have enough accuracy without additional sensors. Of course, sensors can also be placed in the environment to track the subject's progress around the track. In his dissertation, Feldman provides another alternative where he uses several SICK laser scanners to localize players during a basketball game (Feldman 2008). By aiming these scanners at about ankle height, the researcher can get precise information as to where each leg is during each second of the experiment. While precise subject tracking is now feasible, the researcher must weigh the cost in money, set-up, and maintenance of these systems compared to the expected benefit for the experiment.

9.3.1   READING ON-THE-GO STUDY

An early track study that we performed was to investigate the comprehension of text read by participants while in motion (Vadas et. al 2006). Because reading text is a fundamental task performed on mobile devices, we wanted to explore the trade-offs between visual and auditory interfaces for reading. Because reading is a head-down task, we opted to use a taped track on the floor, weaving between obstacles such as desks and chairs (Figure 9.4).

Participants in the study were requested to walk around the track while either reading or listening via text-to-speech software to short passages. After reading or listening to each passage, they used a handheld input device to respond to questions about the passage. A palmtop computer was used to control the study and to record data. On the palmtop, we recorded data such as reading time and response accuracy. One of the researchers in the study counted the number of laps that were made by the participant during each trial, and also kept track (using a handheld counter) of how many times the participant stepped outside of the path. The average walking speed over all of the trials was also computed. We also collected baseline walking performance where the participants did not use any technology.

Given the data collected, we were able to compute statistics about comprehension in conditions approximating real life with the user "on-the-go." In everyday life, a person reading while in motion must perform many of the same tasks as were encountered by our participants, such as obstacle avoidance and navigation. For reading text, the head-down nature of the walking task was more appropriate than a head-up flag track and thus we achieved reasonable ecological validity.

9.3.2 QUICKDRAW STUDY

We conducted a study similar to the Reading on-the-Go experiment in order to determine how much time is required to access a mobile device while walking or standing (Ashbrook et al. 2008). The study involved participants walking around a track or standing stationary while responding to occasional alarms from a mobile device. At semi-random intervals, the device would make a loud noise and vibrate, and participants had to retrieve the device and perform a sequence of actions on the touchscreen. We measured the amount of time it took participants to retrieve the device, to get it into a ready position, and to respond to the alert. Participants wore the device in their pants pocket, in a holster on the band of the pants, and affixed to the wrist (Figure 9.7). These conditions were chosen to represent a reasonable subset of the on-body placements we might expect for several types of wearable devices.

Because the experimental procedure dictated that, in between responses to the alarm, participants had no other task than walking the track (in the mobile condition) or standing still (in the stationary condition), we used a flag-delimited track. This decision made walking more natural but meant we did not get the same statistics as we could with a floor track, such as the number of times the user steps off of the track. The sole walking statistics we were able to record were the average time taken per lap in each condition and the number of times the participant slowed or stopped walking while responding to an alarm.

In this study, the level of ecological validity was higher given the more realistic character of the participant's walking. They walked at a natural rate and responded to alarms in the same fashion as they might respond to a ringing phone. Internal validity, however, was lessened, due to our inability to control for individual factors such as type of clothing. It is clear that there will be a difference in access time for a pocketed device if one is wearing tight pants as opposed to loose pants; however, this is a difficult property to measure. We considered various techniques to measure tightness of pockets, but realized that participants already accustomed to wearing tighter pants could be more facile at retrieving objects from their pockets, which might counteract delays due to tightness.

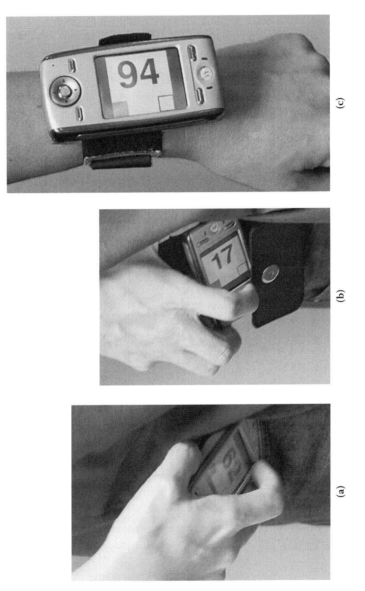

FIGURE 9.7 A mobile device stored (a) in the pocket, (b) in a belt-mounted holster, and (c) on the wrist.

9.3.3   MULTITOUCH TEXTILE INTERFACES

Recently, we have begun to investigate textile interfaces for the control of mobile devices (see Figure 9.8). Specifically, we are investigating "gropability"—the ability of the user to access and control a textile interface while visually distracted or on the go. For example, joggers and skiers often wear MP3 players. Yet the controls on these devices are often small and intricate, requiring users to stop their physical movement and look at the device to control it. We hope to address these problems through an embroidered multitouch textile interface.

The idea behind embroidered multitouch interfaces is to use raised, touch-sensitive embroidery to haptically (rather than visually) guide the user to the buttons as well as make it easy for the user to select the appropriate button without interrupting a primary task. Our prototype uses multitouch to activate the buttons to prevent accidental presses and to allow the user to feel for the desired button before activating it. In order to study the efficacy of this interface, we had a number of options. One option was to perform a stationary, baseline study in order to determine the best-case performance we can expect from users. We could try to improve on a baseline study and add ecological validity by blindfolding users, or making them perform a distracting task on the computer, to simulate the dual-task nature of natural use. For this particular situation, however, we believe that a track-based study will give us the best data.

One of our goals was to create an interface that can be used with little or no visual attention. How can such a system be tested? What should the interface be compared to? We chose to compare our four-button multitouch interface to a three-button single touch interface. The interfaces were identical in design with the exception of the fourth button. To make a selection on the single touch interface, the user feels for the buttons and simply presses the desired one. Because the user accidentally hits several buttons before finding the appropriate one, the system only registers a button

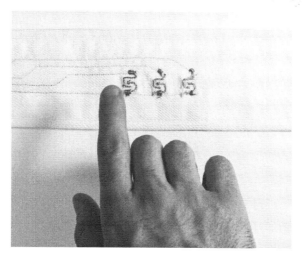

FIGURE 9.8   Textile interface for digital audio player control.

press when the user touches it, and it alone, for more than 2 seconds. Our system plays an audio "beep" when it registers the button press to alert the user that their input has been registered. Since this press-and-hold interface is similar to textile interfaces currently in the literature and to systems currently on the market for ski jackets, the study provides some ecological validity.

9.4   MOBILE IN-THE-WILD STUDIES

Moving to the far end of the ecological validity spectrum are studies conducted "in the wild." With this type of study, the researcher loses most of his or her ability to control variables that may impact study results; however, at the same time, these studies can often provide mechanisms for obtaining information on how mobile and wearable technologies might be used in everyday settings.

One of the most important challenges with these types of studies centers around collecting data. Given the real-world nature of these studies, it is often infeasible for a researcher to directly observe participant behavior during a study conducted in the wild; or if direct observation is possible, it may be limited in scope or duration. Likewise, all data collection must be done on the wearable device and either stored for later analysis, or buffered and uploaded over a network with sufficient coverage such as a 3G+ data link of a mobile phone network.

There are a few methodological approaches for addressing these issues. One common approach for collecting data for studies in the wild relies upon participant self-reporting. For example, the research may use a diary study design whereby participants are asked to record a subset of significant events in a diary that can later be analyzed by the researcher. This technique is obviously limited in fidelity and can be problematic since it totally depends on the participant's compliance. A slightly more structured technique is to use experience sampling. Here, a device such as a mobile phone periodically interrupts the user to elicit responses to a few predefined questions. The interruption provides a cue for the participant, prompting them to record some information for the study and therefore does not totally depend upon the user remembering to record salient events as with diary studies. However, experience sampling is also limited in that it can only be used to capture a small slice of the participant's total experience.

One method we have successfully used for collecting more detail about the use of wearable systems in the wild is the use of mobile capture systems. These systems augment a wearable or mobile computer with additional software to log user activity. Furthermore, we often also utilize video recorders: one video recorder may be connected to a camera mounted on the user facing forward to capture the user's context, and another video recorder may point down to capture the user's hands or the computer's display. Finally, software on the system under evaluation can log system-level user interface events (such as key presses and mouse movement). This approach is quite heavyweight, requiring a great deal of additional hardware, and may only run for a few hours due to battery limitations. However, it also provides a significant amount of data that can be captured while the user performs tasks while out of the laboratory and can be used to get a broad understanding of a wearable computer user's experience.

During an evaluation to produce a case study of an expert wearable computer user, we employed this type of capture technique (Lyons and Starner 2001). For this study, our goals were to catch everyday usage of the computer over long periods of time in the user's "natural" setting. For this evaluation, we created a lightweight capture system that periodically captured screenshots from the user's display and stored them on the wearable's hard drive. These screenshots were then periodically reviewed with the user during focused interview sessions to collectively interpret and reconstruct how the wearable was being used and in what context.

What metrics can be used when the user is "in the wild"? Most of the metrics described in previous sections could, in theory, be captured in the wild depending on the amount of hardware participants are willing to carry. With sufficient resources, GPS, inertial sensors, cameras, and microphones could be combined to capture the participant's location, limb movement, speech, and environment. Yet equipping participants with a great deal of very expensive hardware and expecting them to maintain it is unreasonable. While improvements in technology will reduce many of the current barriers in creating wearable sensor packs, for now in-the-wild experiments should likely focus on details that are difficult or impossible to capture in more controlled conditions.

As stated before, a simple, yet important, statistic to record in the wild is usage. For example, how often do subjects use a new Bluetooth-enabled textile interface to control their phone versus the normal interface? How long are these interactions compared to current interactions? Such data can often be recorded easily as part of a study. If subjects are willing to wear a camera to record the environment in which the interaction takes place, the researcher can interview them after each day or week to get a better sense of the interface's usage, as described above. Interestingly, studies have shown that the NASA TLX can be used successfully after a long delay, with subjects reviewing their video after a series of interactions took place to remind them of their cognitive load at the time (Hart and Staveland 1988).

9.4.1 PRIVACY

One key challenge with using capture as a data collection technique is privacy. Care must be taken by the researcher to avoid violating not only the participant's privacy but also the privacy of those around the participant. Some sensors, such as accelerometers, have low potential to violate privacy: from the perspective of the participant, there is little specific information that can be gained, and the data cannot be used directly to discover the identity of the participant. From the perspective of other people, the accelerometer has no ability to sense them, so poses no threat to privacy. Sensors that have output data that is more understandable to humans are by their very nature more of a threat to privacy, and their use must be carefully considered. GPS receivers reveal where a person has traveled, and such data is difficult to anonymize except by taking it out of the context of a map. Microphones can sometimes be used in a privacy-preserving manner by using noise-canceling microphones; designed to pick up only a person's speech, they usually can filter out background noise, including the speech of other people in the vicinity. One can also discard the speech and instead only record computed features of the audio data.

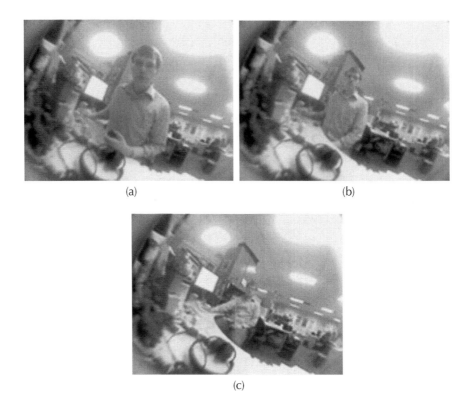

(a)   (b)

(c)

FIGURE 9.9   Three images captured with a fisheye lens. Image (a) was captured at a distance of 0.5 m; image (b) was captured at 1 m; and image (c) was captured at 1.5 m.

Video can be an especially challenging sensor when it comes to privacy. Special considerations must be made to prevent the camera from picking up personal moments, such as restroom use, and to avoid recording people in non-public situations. Local laws must be taken into account, but in the United States, for example, it is only when people have a "reasonable expectation of privacy" that they must not be recorded. One method to avoid recording identifiable video of others, while still maintaining enough clarity that the video can be used for context, is to put a fisheye lens onto the camera and record at lower resolution. Figure 9.9 illustrates this concept, showing images captured at 160×120 resolution at three distances. Unless someone comes unusually close to the researcher, faces in the surroundings do not have enough resolution for identification.

9.5   CONCLUSIONS

Although challenging, the evaluation of wearable computing systems is necessary to effectively perform research in this area. As technology progresses, body-worn technology will only become more and more integrated into everyday life. Mobile phones are now full-fledged computational devices that are worn on the body, and clothing-integrated technology is progressing by leaps and bounds. The techniques outlined in this chapter should serve as a starting point for the interested researcher to design studies appropriate to the technology under consideration.

ACKNOWLEDGMENTS

The research and techniques discussed in this chapter were funded, in part, by grants from ETRI, NSF, and NIDRR's Wireless RERC. This material is based upon work supported by the National Science Foundation under grant number 0812281. Any opinions, findings, and conclusions or recommendations expressed in this material are those of the author(s) and do not necessarily reflect the views of the National Science Foundation. The Rehabilitation Engineering Research Center for Wireless Technologies is sponsored by the National Institute on Disability and Rehabilitation Research (NIDRR) of the U.S. Department of Education under grant number H133E060061. The opinions contained in this chapter are those of the author and do not necessarily reflect those of the U.S. Department of Education or NIDRR.

REFERENCES

Ashbrook, D., J. Clawson, K. Lyons, N. Patel, and T. Starner. 2008. Quickdraw: The impact of mobility and on-body placement on device access time. *Proceedings of the Conference on Human Factors in Computing Systems* (CHI), Florence, Italy.

Ashbrook, D., K. Lyons, and T. Starner. 2008. An investigation into round touchscreen wristwatch interaction. *Proceedings of International Conference on Human-Computer Interaction with Mobile Devices and Services* (MobileHCI), Amsterdam, The Netherlands.

Barnard, L., J. S. Yi, J. A. Jacko, and A. Sears. 2005. An empirical comparison of use-in-motion evaluation scenarios for mobile computing devices. *International Journal of Human-Computer Studies (IJHCS)*, 62 (4): 487–520.

Beauregard, S. 2009. Infrastructureless pedestrian positioning. PhD thesis, Universitaet Bremen.

Clarkson, E., J. Clawson, K. Lyons, and T. Starner. 2005. An empirical study of typing rates on mini-QWERTY keyboards. *Proceedings of the Conference on Human Factors in Computing Systems* (CHI), Portland, OR.

Dix, A. J., J. E. Finlay, G. D. Abowd, and R. Beale. 2004. *Human-computer interaction* (3rd Ed.). New Jersey: Prentice Hall International.

Feldman, A. 2008. Using observations to recognize the behavior of interacting multi-agent systems. PhD thesis, Georgia Institute of Technology, Atlanta.

Hart, S. G., and L. E. Staveland. 1988. Development of a multi-dimensional workload rating scale: Results of empirical and theoretical research. In *Human mental workload,* eds. A. Hancock and N. Meshkat, 139–83. Netherlands: Elsevier.

Hudson, J., J. Christensen, W. Kellogg, and T. Erickson. 2002. I'd be overwhelmed, but it's just one more thing to do: Availability and interruption in research management. *Proceedings of the Conference on Human Factors in Computing Systems*, Minneapolis, Minnesota.

Intille, S., J. Rondoni, C. Kukla, I. Iacono, and L. Bao. 2003. A context-aware experience sampling tool. *Proceedings of the Conference on Human Factors and Computing Systems* (CHI), Ft. Lauderdale, FL.

Lyons, K., D. Plaisted, and T. Starner. 2004. Expert chording text entry on the Twiddler one-handed keyboard. *Proceedings of International Symposium on Wearable Computing* (ISWC), Arlington, VA.

Lyons, K., and T. Starner. 2001. Mobile capture for wearable computer usability testing. *Proceedings of IEEE International Symposium on Wearable Computers* (ISWC), Zurich, Switzerland.

Starner, T., C. Snoeck, B. Wong, and R. M. McGuire. 2004. Use of mobile appointment scheduling devices. *Proceedings of the Conference on Human Factors in Computing Systems* (CHI), Vienna, Austria.

Vadas, K., N. Patel, K. Lyons, T. Starner, and J. Jacko. 2006. Reading on-the-go: A comparison of audio and hand-held displays. *Proceedings of the ACM International Conference on Human-Computer Interaction with Mobile Devices and Services* (MobileHCI), Espoo, Finland.

10 Fundamentals of and Requirements for Solar Cells and Photovoltaic Textiles

Jong-Hyeok Jeon and Gilsoo Cho

CONTENTS

10.1 SOLAR CELLS

A solar cell is an energy generator that converts solar energy to electrical energy. Obviously, solar energy can also produce heat, so-called thermal energy, when photons (particles of light) strike the atoms that make up substances such as a human body, soil, buildings, water, and so on. That is, solar energy can produce electrical energy or thermal energy depending on the power generating substance we choose (i.e., electricity or heat). A solar cell is also called a photovoltaic cell since it produces electricity directly from light. The word photo means light and the word voltaic originated from the name of an electrical engineer, Alessandro Volta (1745–1827), which the famous electrical terminology, volt, is from. That is, photovoltaic means "light electricity."

10.1.1 DEVELOPMENT OF SOLAR CELL

The beginning of solar cell development originated with the French physicist, Antoine-Cesar Becquerel, in 1839. He discovered the photovoltaic effect when he was experimenting with a solid electrode in an electrolyte solution. What he saw was a voltage developing when light fell upon the electrode, which was the first finding of photovoltaic phenomena. In 1883, an American researcher, Charles Fritts, invented the first-ever working solar cell by forming a junction using the semiconductor material, selenium, and a very thin layer of gold. However, the power conversion efficiency, or conversion efficiency, was only around 1% and couldn't be used in real applications, even though it was the first invention of the solar cell. Power conversion efficiency can be defined as following:

$$\eta \equiv \frac{P_{generated}}{P_{in}}$$

where $P_{generated}$ is the generated power from solar cells and P_{in} is the total solar power density incident on the solar cell. Therefore, 100% conversion efficiency means solar cells are absorbing all the light shining on the devices and converting this light to electricity completely, which is an ideal case, though is not actually possible. Another American researcher from AT&T Bell Labs, Russell Ohl, made the first silicon solar cell but still couldn't improve the energy conversion efficiency of the solar cell devices higher than 1% due to the lack of semiconductor technology.

Since the first semiconductor p-n junction was invented in 1948, there has been a tremendous breakthrough in solar cell technology, utilizing many different kinds of semiconductor materials such as Si, GaAs, CdTe, and so on. Innovation of semiconductor device technology finally accelerated solar cell development. In 1954, three American researchers, Gerald Pearson, Calvin Fuller, and Daryl Chapin, invented a silicon-based solar cell that achieved around 6% conversion efficiency and opened a new world for practical solar cell technology. Today's best silicon solar cells are over 20% efficient, with commercial averages over 15%.

Currently, solar cells are being researched and developed with polymer materials since this technology is expected to lower the cost of solar cell devices to the extreme even though the technology has not matured yet. First-generation solar cells were made for people deep in the mountains, the military, and outer space satellites, where there is no fundamental energy source such as wall plug electricity, oil, coal, and so on. This has been why solar cells nowadays can be the most powerful long-term power supply for satellites and space vehicles. However, nowadays, with the growing motivation of "Green Energy" and high oil prices, solar cells are now becoming more and more popular for ordinary people. Considering that conventional energy resources, such as fossil fuels, will be exhausted in the near future with increasing demand of worldwide energy, we must find and develop alternative energy sources. From this viewpoint, solar energy is the first long-term energy source for human beings, and solar cells are considered as a major candidate for harnessing this resource. Large-scale solar power plants are being built with hundreds of megawatts power levels. Also, various kinds of solar cells are being used in applications on building materials, airplanes, automobiles, clothing, and so on.

10.1.2   SOLAR SPECTRUM

The sun emits solar energy primarily as electromagnetic radiation with a spectrum ranging from infrared to ultraviolet (200 ~ 3000 nm light wavelength) with the amount of $4 \times 10^{26'}$ J every second, that is, 4×10^{26} watts. In order to make a good solar cell, understanding of the solar spectrum is of great importance since the energy is from the sun. The spectral distribution of solar energy on the Earth is shown in Figure 10.1. The AM0 (air mass zero) curve represents the measured solar spectral power density right outside the Earth's atmosphere and is very important for satellite applications. On the other hand, the AM1.5 curve represents the average spectral power density at the Earth's surface in the United States. The AM0 curve gives the total area of around 135 mW/cm^2 in the graph. However, the AM1.5 curve gives the total area of around 100 mW/cm^2, which means solar cells can generate the power of 100 mW with a 1×1 cm^2 size solar cell assuming 100% solar cell conversion efficiency on the Earth's surface. In other words, for example, if we have a total area of 20 m^2 of solar cell arrays with 25% conversion efficiency on the roof, then we can get electricity of

Solar cell generated electricity = 0.25×100 mW/cm$^2 \times 20$ m$^2 = 5$ kW.

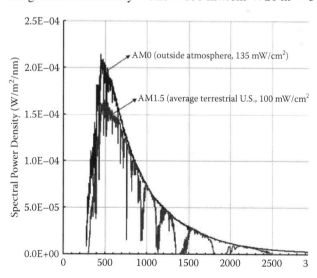

FIGURE 10.1   Solar spectrum.

Assuming a solar cell can take the sunlight for about 10 hours per day on average, it can produce 50 kW-hours per day and 1,500 kW-hours per month, which is higher than the United States average household electricity consumption rate of 920 kW-hours per month in 2006. Considering the average price of residential electricity in the United States in 2006 was $0.11 per kW-hours, the solar cell can produce 1,500 × 0.11 = $165 per month and about $2,000 per year in electricity.

Obviously, AM0 curve gives more solar energy than AM1.5 because the atmosphere attenuates the sunlight before it reaches the earth's surface. Our main concern for typical solar cell applications is with AM1.5. In the plot, we can observe the rainbow light spectrum region (light wavelength of 400 ~ 750 nm region) has the main part of this AM1.5 solar spectrum. Therefore, we can see that the light spectrum in this region is the most important portion of the solar spectrum to make a good solar cell. One notable point is that there exists no material that can absorb all the sunlight in this rainbow region, which is still a critical issue for high-efficiency solar cell research. That is, one material can only absorb a certain range of sunlight efficiently and that range is very narrow (only about a few hundred nm). For this reason, many researchers are investigating the possibility of fabricating two or three different materials-based solar cell structures. That is, they divide the solar spectrum region by two or three different regions and select proper materials for the corresponding spectrum region, and finally they can collect as much sunlight input as possible from two or three different materials. Sometimes, they even try several different materials for this kind of solar cell structure. However, this multi-material solar cell approach will be expensive in comparison to its higher conversion efficiency. Currently this kind of technology is being used in military and satellite applications where performance issues dominate cost.

10.1.3 Trend of the Solar Cell Structure and Its Conversion Efficiency

Over the last 30 years, with the growing motivation of making a high efficiency and low cost solar cell, there have been remarkable breakthroughs in solar cell conversion efficiency.

Currently, there are about three different schemes for solar cell research and development (Figure 10.2):

- Single-junction semiconductor solar cell approach: This approach takes advantage of the simplicity of the structure and hence can reduce the solar cell fabrication cost but is limited by lower conversion efficiency. As mentioned earlier, there does not exist any semiconductor material that can absorb the entire solar spectrum. One material can only absorb a certain narrow range among the whole solar spectrum. For this reason, single-junction structure solar cells have only one region that can be absorbed efficiently in the solar spectrum range. However, its simple structure can be a strong advantage from the viewpoint of cost. Currently, this approach is being taken as a solution for mass production of solar cells due to its relatively low cost and intermediate conversion efficiency. Eventually, the conversion efficiency will be fixed and there will not be room for improvement of the efficiency.

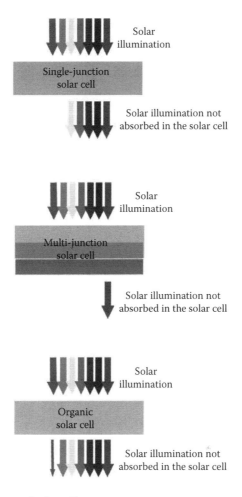

FIGURE 10.2   Schemes of solar cell structure.

- Multi-junction semiconductor solar cell approach: This approach takes advantage of a multi-material solar cell structure to get higher conversion efficiency with somewhat more complicated structures. Since each junction with a different material of this multi-junction structure can absorb a specific corresponding solar spectrum, the whole solar cell can absorb a wider range of the solar spectrum and hence generate relatively high conversion efficiency. In spite of its higher cost to produce, this scheme has a very important role in the applications of satellites, space vehicles, military equipment, and so on. Whenever there is limited space or no other energy source exists, this approach has the most efficient way to collect solar energy.

- Organic-material-based solar cell approach: This is a relatively novel technology since it utilizes organic (polymer) materials, which has advantages of being lightweight, very low cost, and flexible. However, conversion efficiency is much lower than the previous two schemes. Once this technology is able to make notable breakthroughs in reaching higher efficiencies then it could dominate solar cell research directions. However, organic solar cells also suffer from environmental degradation. Good protective coatings are still being developed.

The following are current solar cell conversion efficiency records of the three schemes so far:

- Multi-junction solar cell: 42%
- Single-junction solar cell: 25%
- Organic solar cell: 5.5%

10.1.4 SOLAR CELL BASICS

In this section, we will investigate the fundamental nature of semiconductor p-n junction diodes because this is at the root of all semiconductor-based solar cells. Electrically, there are two different kinds of semiconductor materials: p- and n-type. The p-type semiconductors have "holes" as electrical charge carriers that have a positive sign, and the n-type semiconductors have "electrons" with a negative sign. All the charge carriers are free to move around inside the p- and n-type materials. Each type of semiconductor is electrically "neutral" because all the atoms in each material are electrically neutral. Once an atom loses one of its electrons (holes), it can compensate another electron (hole) from neighbor atoms so the whole material can maintain the electrical neutrality. When a p-type semiconductor makes a chemical bond with an n-type semiconductor, as shown in Figure 10.3, we can say the two different types of semiconductor materials form a p-n junction and have some unique properties. Again, the p-type material has positive majority charge carriers, called holes, which are free to move around inside the crystal lattice, and the n-type material has negative majority carriers, called electrons. Electrons in the n-type material diffuse across the junction, combining with holes in p-type material at the interface between two different types of materials. The region of the p-type material near the junction takes on a net negative charge because of the electrons attracted. Since electrons departed the n-type region, it takes on a localized positive charge. The thin layer of the crystal lattice between these charges has been depleted of majority carriers and is known as the depletion region. Essentially, it becomes a nonconductive intrinsic semiconductor material. In effect, there is an insulator separating the conductive p- and n-type regions, which is electrically in equilibrium. This is also called zero-bias equilibrium because there is no external voltage bias on the p-n junction.

The separation of charges by the depletion region at the p-n junction constitutes a potential barrier that works against carrier transport and hence there is no more charge transport (no current). This potential barrier can be lowered or even enhanced by an external voltage source. Once the barrier height is lowered by forward bias (i.e., positive bias for p-type and negative for n-type), then the depletion region thickness decreases. Now, the carriers (electrons and holes) can move towards the junction interface easily, and once they pass the junction borderline of the depletion region,

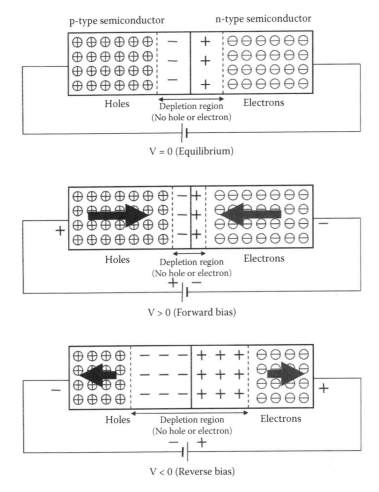

p-type semiconductor   n-type semiconductor

Holes   Depletion region   Electrons
(No hole or electron)

V = 0 (Equilibrium)

Holes   Depletion region   Electrons
(No hole or electron)

V > 0 (Forward bias)

Holes   Depletion region   Electrons
(No hole or electron)

V < 0 (Reverse bias)

FIGURE 10.3 p-n junction diagrams without light illumination.

they actually move electrical charges and we say there is current flowing. On the other hand, when the barrier height is increased by applying a reverse bias (i.e., positive bias for n-type and negative for p-type), then the depletion region thickness increases, which results in total blockage of any charge transport through the depletion region and there is no current flowing at all. Finally, we have the current-voltage (IV) characteristics curve for a p-n junction semiconductor device in Figure 10.4. The formation of the junction and potential barrier happens during the manufacturing process. The magnitude of the potential barrier is a unique property of each semiconductor material.

So far, we have investigated the intrinsic property of p-n junction at dark condition (i.e., no light illumination). Since solar cells are nothing but p-n junction devices under light illumination, it is very important to understand the p-n junction characteristics when there is light illumination. When light is incident on semiconductor material, light particles, so-called photons, are absorbed in the semiconductor

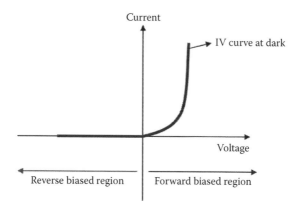

FIGURE 10.4 Typical current-voltage (IV) characteristics of p-n junction without light illumination.

material and generate electron-hole-pairs (EHPs). This process happens everywhere in all types of semiconductor materials. One notable point is that only generated electrons in p-type semiconductors and generated holes in n-type semiconductors will move to the junction direction because of the potential in the depletion region, which was pointed out earlier. Actually, this transport has reversed direction to the forward biased characteristics of p-n junction devices and hence shifts the IV curve to the negative y-axis direction by the amount of photo-generated current, which is described in Figure 10.5. Since this photo-generated carrier transport also happens with reverse biased p-n junctions, the IV curve will also be shifted to the negative y-axis direction by the same amount. Hence, the IV curve at both forward and reverse biased region will be shifted in parallel by a certain constant amount of photo-generated current in the negative y-axis direction.

Figure 10.6 represents typical p-n junction current-voltage (IV) characteristics with and without light illumination. As shown in the plot, when there is light input to solar cells, we can see the dark IV curve shifts to the negative y-direction, which is the key point of power generation by solar cells. That is, in the fourth quadrant,

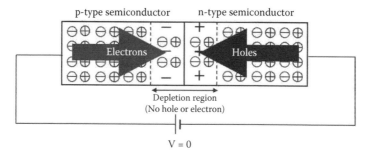

FIGURE 10.5 p-n junction diagram with light illumination. (EHPs generated by photons are distributed everywhere in the semiconductor material. Only carriers generated by photons are shown for easier understanding.)

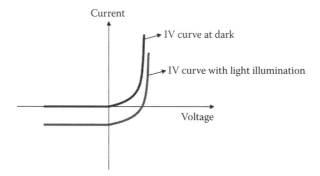

FIGURE 10.6   Typical solar cell current–voltage (IV) characteristics in dark environment and with light illumination.

the multiplication of the current and voltage (in other words, the power) has negative sign, which means power is "generated" instead of "consumed" by solar cells. In the first quadrant, the signs of the voltage and current are both positive and, in the third quadrant, the signs are both negative. In either case, the multiplication of current and voltage, which means electrical power in watts, has a positive sign and hence we need to "consume" electricity to turn on all the home electronics, computers, MP3 players, and so on, since all electronic devices except solar cells are being operated in the first or third quadrants with any kind of circuit. On the other hand, since solar cells are being operated in the third quadrant, they actually "generate" rather than "consume" the electricity. For this reason, solar cells can be "energy generators" as long as there is solar illumination on the devices.

For easier understanding, since the fourth quadrant of the graph is of prime interest for solar cell analysis, it is common practice to draw the fourth quadrant upside down, which is illustrated in Figure 10.7. In the plot, the following are parameters of interest:

V_{OC}: The open circuit voltage (the maximum voltage obtainable by solar cells).

I_{SC}: The short circuit current (the maximum current obtainable by solar cells).

V_m, I_m: The actual operating voltage and current in solar cell application with maximum $I_m \times V_m$ rectangle area, which is P_{max}, maximum power level that can be driven by the solar cell.

In the plot, we see that we need to make the $I_m \times V_m$ rectangular in shape and as large as possible to make good solar cells. This means that V_{OC} and/or I_{SC} should be increased. Note that the area of $I_m \times V_m$ rectangle has the units of electrical power in watts.

Another parameter for solar cells is called the fill factor, FF, and is defined as

$$FF \equiv \frac{P_{max}}{I_{SC}V_{OC}} = \frac{I_m V_m}{I_{SC}V_{OC}}$$

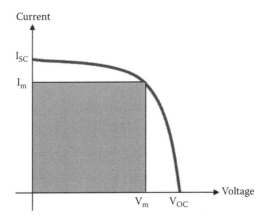

FIGURE 10.7 Inverted plots of the fourth quadrant IV characteristics of solar cells.

Obviously, improving the fill factor is also a very important task for better solar cell performance.

Since solar cells are electricity generators, it is not difficult to apply these devices in normal life. They can be directly connected to any home electronics such as TVs or computers as described in Figure 10.8. Or, when we are not using any electricity or are using less electricity than the solar cells are generating, we can sell the extra electricity to the electric power company. In this sense, solar cells can even generate money when we are not using the electricity from them, which could be another big merit of this device.

10.1.5   FACTORS TO IMPROVE SOLAR CELL PERFORMANCE

Figure 10.9 illustrates a typical solar cell structure. Even though there exist various kinds of solar cell structures, most solar cells have the basic structure shown in the diagram. Now, we will investigate some considerations for improving solar cell performance.

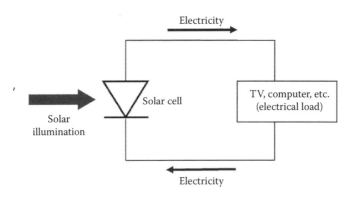

FIGURE 10.8   Solar cell applications.

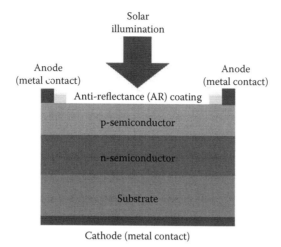

FIGURE 10.9   Typical solar cell structures (single junction).

10.1.5.1   Carrier Lifetime

As mentioned earlier, we have two different kinds of charge carriers that are generated by a solar cell device (i.e., electrons and holes) and each of them must pass the
junction depletion region and reach the other side to have electrical current flowing.
As it turns out, these photo-generated carriers have their own lifetimes depending
on the solar cell device structure (such as the thickness and doping concentration of
p and n semiconductor layers) and the material quality. In other words, the photo-
generated carriers could disappear before they reach the other side of the junction
since electrons and holes can recombine together very easily. In this case, there will
be no photo-generated current flowing through the device even with solar illumination. The point is knowing whether the carriers can reach at the other side of junction
before they recombine together. This is directly related with the optimization of the
solar cell device structure and the semiconductor material quality. In fact, this is
the major issue in solar cell research. Carrier lifetime plays an important role in the
open-circuit voltage (V_{OC}) and the short-circuit current (I_{SC})

10.1.5.2   Metal Contact

Since a solar cell is one type of a semiconductor device, it needs electrodes such
as anodes and cathodes to collect the charges that are generated. However, whenever a metal makes contact with a semiconductor material, there exists an electrical
resistance between the metal and the semiconductor. Higher contact resistance will
hamper the solar cell performance, and therefore a good metal contact is another
essential factor to make good solar cells. This metal contact mainly affects the fill
factor characteristics of solar cells. Another consideration related to this metal contact is that, as shown in the plot, the top-side metal contacts (anodes) are blocking
the solar illumination and end up with loss of solar light input. The trade-off is when
we have narrow metal contacts and are able to get more solar illumination, the metal
contact resistance goes up and results in the loss of the fill factor. Therefore, the

metal contact design should be optimized for the best performance of the device. In order to avoid this issue, transparent metals such as indium-tin-oxide (ITO) are also being researched to be applied in solar cell engineering.

10.1.5.3 Anti-Reflectance Coating

When a light is incident on any semiconductor material, some portion of that light will be reflected by the surface due to the difference in material density. Anti-reflectance (AR) coatings can considerably decrease this reflectance and hence make the light penetrate the semiconductor material more readily and be absorbed. This coating is usually fabricated with an oxidized semiconductor material and has different material formation depending on the specific solar cell material. Obviously, this factor will affect the open-circuit voltage (V_{OC}) and the short-circuit current (I_{SC}) of the solar cell characteristics.

10.1.6 FUTURE TREND

So far, we have investigated the solar cell basics and some trends in the solar cell field. However, no single solar cell scheme is dominating the market or even academic society yet because solar cell technology has not fully matured in terms of cost-effectiveness. Recently, a new approach with an optical concentrator (lens) was proposed and is expected to reduce the solar cell cost significantly, as described in Figure 10.10. This concentrator can focus the solar illumination into the solar cell devices, which can save the solar cell material expense because the concentrator is less expensive than an equivalent area of solar cell. Additionally, this idea encourages the development of relatively high-cost, high-efficiency solar cell structure, that is, multi-junction solar cells. Many worldwide companies are still developing low-cost single-junction solar cells using specialized manufacturing techniques in order to reduce the cost even further. Even today, this is being commercialized and brought to regular markets. However, in order to produce solar cell–generated electricity at a cost comparable to traditional energy technologies such as nuclear power, the single-junction solar cell scheme must reduce its costs even further. It can, though, still be commercialized partially for specific purposes, such as where it is relatively hard to find traditional energy sources. As for organic solar cells, even though this is really promising due to its very low cost merit, more academic breakthroughs are needed at the research level to increase its low conversion efficiency.

10.2 PHOTOVOLTAIC TEXTILES

If appropriate research is quickly and successfully undertaken, the general public and professionals may ordinarily use electronic devices powered by solar energy in most countries by 2025. Such a situation is highly desirable because sustainable development is a great concern worldwide and because the emerging ubiquitous environments will become unmanageable if people must manually recharge or change the batteries of all involved artifacts. In theory, buildings are easily powered by solar cells thanks to their relatively large surface, but not the objects

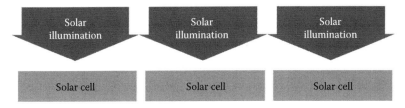

(a) Solar cell array without optical concentrators

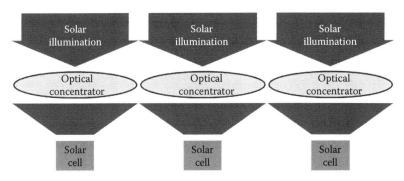

(b) Solar cell array with optical concentrators

FIGURE 10.10   Solar cell arrays with and without optical concentrators. (Note that concentrator scheme can save a huge amount of solar cell material.)

placed inside (e.g., toys) or mobile devices (e.g., cellular phones) due to their size, shape, and sometimes required flexibility or tensility. Efforts are already ongoing to develop flexible cells—with approaches like dye-sensitized (Lee and Won 2008) and polymer-based (Kim 2008) solar cells—but other properties will be requested as well. As an alternative way to harvest solar energy, we thus propose to take a completely different path by inventing *photovoltaic textiles*: textiles that convert sunlight into electricity.

Figure 10.11 illustrates how photovoltaic textiles could contribute to everyday life, with solar rays stimulating materials in textiles to send electricity to standard devices (e.g., cellular phones), enhanced clothes (e.g., coats incorporating music players), or enhanced accessories (e.g., bags storing energy). Such textiles could support various situations by, e.g., covering remote-controlled cars, providing flexible pads to plug into mobile personal digital assistances to supply solar energy while relaxing with the PDAs on terraces, and composing smart clothes. For this, however, the textile materials would need to be much more flexible, more resistant to torsion, and more tensile than existing photovoltaic films.

Hereafter, we establish milestones for the realization of successful photovoltaic textiles, describe available methods to obtain yarns, and state potential benefits of our ongoing research.

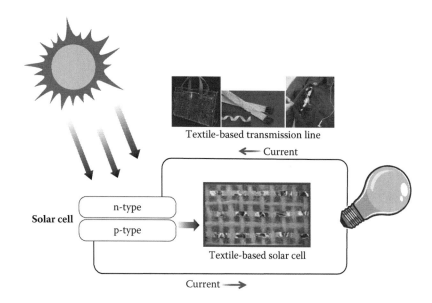

FIGURE 10.11   Lights into solar energy.

10.2.1 MILESTONES FOR THE REALIZATION OF PHOTOVOLTAIC TEXTILES

Photovoltaic textiles would benefit numerous fields, including clothing and transportation, with few constraints regarding the space needed and location/orientation of the equipment. Excluding the requirement analyses, we identify five milestones for the realization of successful photovoltaic textiles, listed below in chronological order.

First, we must identify photovoltaic materials that can be turned into fibers and textiles. These materials may be organic or inorganic, ideally flexible and tensile but resistant to repeated abrasion and/or bending that might occur during everyday use. As a first example, water-soluble and lipo-soluble CIGS (CuInGaSe2) nanoparticles that can be liquefied are appropriate materials to generate so-called *slit yarns* through slit film processes.

Second, we must develop fibers from elastomeric polymer materials using either conventional or electro-spinning processes then provide the photovoltaic properties through coating or electro-less plating. Three critical aims will be to maximize the durability, the energy transmission, and the efficiency of the conversion from sunlight to electricity through structural optimizations (e.g., thickness) of the fibers.

Third, we must create textiles from these photovoltaic fibers, starting with promising standard techniques such as weaving or knitting. The fabrication methods should be evaluated and selected to maximize the flexibility and efficiency of the solar energy harvesting, at least based on theoretical models (e.g., for exposition surface per fiber weight) and experiments in controlled environments (e.g., solar intensity and pollution in a laboratory), and potentially based on real-world outdoor experiments (e.g., in a city and in a forest).

Fourth, we must invent efficient electro-textile interfaces adapted to each textile fabrication method. Micro-interfaces between photovoltaic fibers and macro-interfaces within fabrics/garments should exist for both two-dimensional and three-dimensional connections.

Finally, we should propose advanced prototypes based on clothing and textile self-recharging in everyday life, including solutions working in big cities like New York or Seoul, to validate the appropriateness of our final materials for different families of application. These suggestions may involve textile-based modules powering mobile/wearable devices, or photovoltaic textile kits for bags, accessories, upholstery, or home interiors.

10.2.2 METHODS FOR THE CREATION OF PHOTOVOLTAIC YARNS

Photovoltaic textiles are based on photovoltaic fibers that are converted into yarns through slit yarn, monofilament spinning, or conjugate spinning processes. These yarns are then converted into textile fabrics through weaving, knitting, felting, laminating, embroidering, quilting, or braiding, which potentially influences differently the energy-harvesting efficiency depending on the constituent materials. Unlike solar cells, textiles can allow the capture of energy from any direction whatever their size because they are flexible; multi-layered structures may even improve efficiency per surface unit further.

10.2.2.1 Slit Yarn Process

Organic and inorganic materials that can form ordinary solar films can be turned into slit yarns through the standard *slit yarn process* shown in Figure 10.12. The slit yarns can then be combined into textiles using diverse methods such as weaving, knitting, embroidery, quilting, and braiding (see Figure 10.13). This process is simple, cheap, and reliable but the resulting yarns can only be flat.

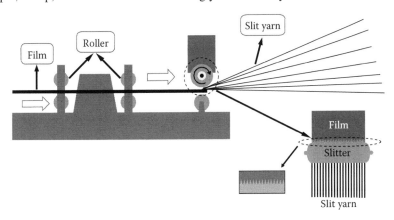

FIGURE 10.12 Manufacturing process for slit yarn.

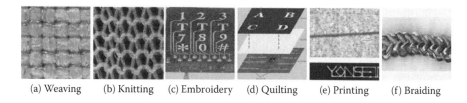

(a) Weaving　　(b) Knitting　　(c) Embroidery　　(d) Quilting　　(e) Printing　　(f) Braiding

FIGURE 10.13　Methods for textile-based photovoltaic cells.

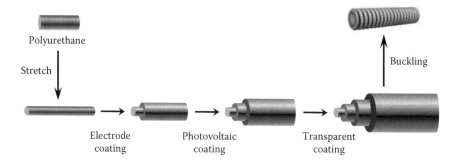

FIGURE 10.14　Coating methods.

10.2.2.2　Monofilament Formation and Coating Methods

Monofilaments using elastomeric polymers such as polyurethane are first obtained using the conventional spinning process. While the monofilaments are stretched, three coatings are applied by, e.g., chemical bath deposition: (1) electrodes, (2) photovoltaic materials, and (3) transparent electrode materials; as shown in Figure 10.14, the filament will form wrinkles after release. This technique ensures that the coating-related properties are maintained even when the filament is stretched again during everyday use.

10.2.2.3　Conjugate Spinning

The *conjugate spinning* process represented in Figure 10.15 is popular in the textile industry to make multifunctional textiles (e.g., anti-static and flame retardant). The electro-spinning process forms conjugate fibers, in our case made of three layers: conductive polymer materials as inner layer, organic semi-conducting materials in the middle, and TiO_2 for protection as outer layer. Afterwards, transparent electrode materials are coated on the surface by chemical bath deposition or electroless plating using nano-scale metallic particles. This advanced process embeds all functions into a single yarn but remains expensive and unreliable for photovoltaic textiles.

10.2.3　Expected Contribution

To our knowledge, the first and only project concerning photovoltaic textiles is the Swiss "Photovoltaic Fibers and Textiles Project" (2004–2008), which focused on

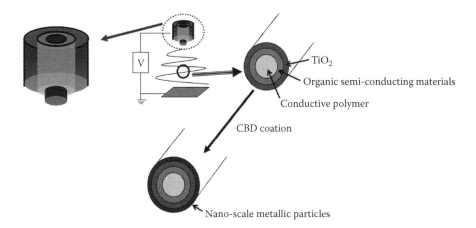

FIGURE 10.15 Conjugate spinning.

(1) improving the physical and mechanical performances of photovoltaic fibers, (2) integrating them into woven textiles, and (3) finding out appropriate inter-connection methods to create useful devices (Konarka 2008). Unfortunately, this project led only to proposals for methods (Morrison 2005), and not to the realization of textiles.

We expect photovoltaic textiles to become a significant technological source for the creation of both active and passive electronic textiles as well as to allow the realization of textile-based active electronic devices. These textiles should contribute to most human environments because fibers serve in both everyday and specialized clothing (e.g., casual, military, and sports), because they are applied as architectural materials, and because they become components of home furnishing or even transportation systems (including cars, ships, and planes). Finally, the use of photovoltaic textiles will allow mobile energy capture and regeneration with lower location constraints.

REFERENCES

Kim, K. 2008. Polymer solar cells. EP&C NEWS. http://www.epnc.co.kr/article/view.asp?article_idx=8480.

Konarka. 2008. Photovoltaic fibers and textiles based on nanotechnology. ARAMIS. http://www.aramis.admin.ch/Default.aspx?page=Texte&projectid=22200.

Lee, D., and C. Won. 2008. A study on the TiO_2 paste for flexible dye sensitized solar cell prepared at low temperature. *Chemistry* 12 (1, May): 189–92.

Morrison, D. 2005. Photovoltaic R&D aims to weave powerful fabrics. Power Electronics Technology. http://powerelectronics.com/news/photovoltaic-weave-fabrics/.

Index

Printed and bound by CPI Group (UK) Ltd, Croydon, CR0 4YY

23/10/2024

01777670-0009